D1253503

IT'S ELEMENTAL

IT'S ELEMENTAL

THE HIDDEN CHEMISTRY IN EVERYTHING

KATE BIBERDORF

Recycling programs for this product may not exist in your area.

ISBN-13: 978-0-7783-8942-2

It's Elemental: The Hidden Chemistry in Everything

This publication contains opinions and ideas of the author. It is intended for informational and educational purposes only. The reader should seek the services of a competent professional for expert assistance or professional advice. Reference to any organization, publication or website does not constitute or imply an endorsement by the author or the publisher. The author and the publisher specifically disclaim any and all liability arising directly or indirectly from the use or application of any information contained in this publication.

This edition published by arrangement with Harlequin Books S.A.

Park Row Books
22 Adelaide St. West, 40th Floor
Toronto, Ontario M5H 4E3, Canada
ParkRowBooks.com
BookClubbish.com

Printed in U.S.A.

For *my* chemistry teacher,

Mrs. Kelli Palsrok

TABLE OF CONTENTS

INTRODUCTION

Nerds like us are allowed to be unironically enthusiastic about stuff.
Nerds are allowed to love stuff,
like jump-up-and-down-in-the-chair-can't-control-yourself love it.
...When people call people nerds, mostly what they're saying is "you like stuff."
—JOHN GREEN

I want to start this book by admitting something.

I am a *nerd* about chemistry.

I'm a chemist, my husband, Josh, is a chemist, and most of our friends are scientists too. (Not all, but nobody's perfect.) I've been known to hold casual conversations about quarks. Josh and I have discussed the parameters of a Nobel Prize–winning experiment on date night and had an intense argument about which element on the periodic table is the best—it's palladium, obviously.

But I know not everybody is like that.

In fact, *most* people aren't.

Chemistry can be hard to understand. Heck, *science* in general can be difficult to comprehend. There are so many terms and rules, and everything can seem so incredibly complex. That's especially true with chemistry because we can't *see* any of it.

In biology, you can dissect a frog.

A teacher can show you the properties of physics, like acceleration, in real life.

But I can't just hand you an atom.

Even my friends and family don't get what I do sometimes. My BFF, Chelsea, is the perfect example. She's super smart, understands science in general and even has a job that's related to chemistry as a jeweler. But Chelsea never "got" what was going on in our high school chemistry classes. While I was enthralled, Chelsea felt simultaneously bored and lost. And when we were sophomores in high school, I simply didn't understand how she felt.

But today, I totally get it. I see students like Chelsea nearly every day.

As a professor at the University of Texas at Austin, I teach a class called Chemistry in Context. It's an introductory course intended for students who will probably never take another science class ever again. Imagine an English major trying to take the easiest science course she can get a C in, and you'll have it.

One year, on the very first day of class, a student asked me a question about quarks, and I ended up in a long tangent about subatomic particles in front of five hundred brand-new freshmen. Some students were trying to frantically take notes while a bunch were just staring at me in various states of shock and fear. Others had resorted to filming me on their phones. Two girls were literally clutching each other.

The whole incident would have been funny if I hadn't freaked out hundreds of students who had decided to give chemistry (and me) a chance. Most of the students had no idea what I was talking about. I might as well have been speaking Klingon. I'm sure this situation only reinforced the myth that science is boring and impossible to understand.

That's because words matter—especially when we talk about science.

When I first got my PhD, I emailed my mom a copy of my dissertation. Minutes later, she called me. Before I could even say hello, I heard her laughing and didn't understand why. Did I send the wrong attachment? Had she just watched a silly cat video? Did she butt-dial me?

Finally she sputtered, "Katie, I don't know what any of these words mean! What's an ass…napthyl?" My mom was laughing so hard that she couldn't say anything else. I was so confused. I had told her what my research was about. Why wasn't she getting it?

Then I opened the document and read the first line:

> The syntheses and catalytic properties of six new 1,2-acenaphthenyl N-heterocyclic carbene-supported palladium(II) catalysts are presented. The acenaphthenyl carbene can be prepared using either mesityl or 1,2-diisopropyl N-aryl substituents.

In that moment, I understood—what my mom read, what my students heard, and what Chelsea had felt. My mom had no idea what I meant by *1,2-acenaphthenyl N-heterocyclic carbene-supported palladium(II) catalysts.*

And frankly, she didn't need to. (In case you're wondering, it's a type of catalyst used in reactions needed to make pharmaceutical drugs.)

Chemistry is so cool, so freaking awesome, but often chemists (including me) talk about the science in a way that loses everybody without a PhD. In this book, I'm going to do the opposite. My mission is to show my mom—and all of you—*why* I am passionate about chemistry. Why it's amazing, why it's super exciting, and why you should love it too.

I promise there will be no discussion of quarks or even a description of the scientific method. But by the time you finish reading, you'll have an understanding of basic chemistry, and you'll see that chemistry is hidden in everything, from the shampoo you use on your hair in the morning to that beautiful sunset at night. It's in the air you breathe, which is something that you literally cannot live without. It's in everything you touch and encounter every single day. And the more you know about it, the more you can appreciate the world we live in.

Just look around yourself right now. Everything you see is matter. All matter is comprised of molecules, and molecules are made of atoms.

The ink on this page is a molecule that has been absorbed into the fibers of the paper, and the glue in the binding is just a fancy molecule that binds to both the paper and the cover. Chemistry is every*where* and in every *thing*.

In the first four chapters, I'll explain what you need to know in order to understand the basics of atoms, molecules, and chemical reactions. You can think of this as Chemistry 101, or a recap of what your teacher was talking about while you were writing notes to your bestie in tenth grade. (BTW, I promise that by the end of this section, you'll finally "get" atoms.)

The second part of the book is about chemistry in everyday life, from the coffee you make in the morning to the wine you drink at night. In between, we'll do all sorts of fun things: bake, clean, cook, work out, even go to the beach. Along the way you'll learn about the chemistry at work in your cell phones, sunscreen, and in fabrics, among other stuff you use every day.

I wrote this book with the hope that you'll not only "get" chemistry but feel excited about it. I hope you'll discover something new and unexpected about the world around you—and that you'll want to share what you learn with your partner, your children, your friends, your colleagues at work…even strangers at happy hour.

Because I strongly believe that we can make the world a better place with a love of science.

Let's get started.

PART ONE

NOT YOUR HIGH SCHOOL CHEMISTRY

1

THE LITTLE THINGS MATTER

The Atom

Chemistry is everywhere, in everything. It's in your cell phone, your body, your clothing, and your favorite cocktail! It explains the way ice melts into water and helps us predict what will happen when we bring two elements, like sodium and chlorine, together (spoiler alert: they make salt).

But what is chemistry, anyway?

The technical definition of chemistry is the study of energy and matter, and how those two things interact with each other. In this context, *matter* refers to anything that exists, whereas *energy* refers to the reactivity of a molecule. (A molecule is one of the super tiny things that matter is made of. More on that later.)

Chemists always want to be able to predict reactivity between two molecules—in other words, predict what will happen when two chemicals or substances come together. So we ask, and try to answer, different questions. Will the chemicals react at room temperature? Will an explosion occur? If we add heat, will that encourage new bonds to form?

In order to answer these sorts of questions, we need to understand the basics of chemistry. This means that we have to go way back because chemistry is an ancient science. Like seriously ancient.

Back in the fifth century, two philosophers named Democritus and Leucippus hypothesized that everything in the world was made up of tiny, indivisible pieces called *atomos*. In a series of essays, these two philosophers described how millions of *atomos* could be joined together to create what we see in the world around us, like a pile of Legos can be used to construct different objects such as a boat or a super cool *Millennium Falcon*.

Though Democritus and Leucippus were completely right—and are today credited with being the first to define the idea of the atom—at the time, their theory wasn't accepted. That's because it contradicted the ideas of two other philosophers at the time, Aristotle and Plato (who were kind of a big deal).

Aristotle and Plato believed that all matter (i.e., everything) in the world was made up of a combination of earth, air, water, and fire. According to this theory, earth is cold and dry, water is cold and wet, air is hot and wet, and fire is hot and dry. And everything in the world could be made from combinations of these four elements. They also believed that every object in the world could transform from earth to air to fire to water and back again. For example, with their theory, when a log is burned, it changes from something cold and dry (earth) into something hot and dry (fire). Once the fire is extinguished, the burnt log turns back into earth because it is now cold and dry.

However, if someone uses water to put out the fire, then the burned log would become a combination of the two elements earth and water. In this example, the burned, wet ashes would take up significantly more space than a just pile of dry ashes. Therefore, Aristotle and Plato were of the opinion that this meant that all matter can get infinitely bigger and smaller, just by changing the combinations.

Democritus hated this idea because he believed there had to be a limit to how small something could become. For example, let's say you divide a loaf of bread in half. Then divide it in half again and again and again. Eventually, Democritus believed that you would run out of bread to divide. When you couldn't divide anymore, he was convinced that last little piece was an individual *atomos*. And Democritus was right!

But once again, it didn't matter because Aristotle was the celebrity philosopher of his time. When Aristotle dismissed the idea of *atomos*, everyone else did too. Unfortunately for us, Aristotle was wrong and humankind spent the next two thousand years incorrectly interpreting the world as some combination of earth, water, air, and fire.

Let that sink in. Two *thousand* years!

It wasn't until the 1600s that someone provided strong enough evidence that could actually challenge Aristotle's theory. A quirky physicist named Robert Boyle loved to perform experiments in order to disprove widely accepted theories. He turned his attention to Aristotle's theory and wrote a book arguing that the world was not made up of earth, water, air, and fire like the Greeks thought.

Instead, Boyle explained that the world was made up of *elements*—small pieces of matter that couldn't be broken into two smaller sections. Sound familiar?

The publication of Boyle's book—aptly named *The Sceptical Chymist* (or *The Skeptical Chemist* in modern English)—started a race to find these small, indivisible pieces called elements. At the time, Boyle believed that common materials like copper and gold were a combination of elements. But shortly after his publication, these materials (and eleven others) were quickly identified and defined as elements.

For example, copper was first used in 9000 BC in the Middle East, but it wasn't until Boyle's book was published that people started to look at it more closely. With the new idea around elements, scientists started to believe that copper was not a combination of elements—but just one element by itself.

The same thing happened with lead, gold, silver...which is how the first thirteen elements were identified. After that, scientists were constantly searching for any evidence that pointed to a new element. This led to the discoveries of phosphorus in 1669, followed by cobalt and platinum in 1735.

Today, we know that an element is exactly what Boyle described: a substance that cannot be broken down further into simpler or smaller substances during a chemical reaction. We also know that elements are made up of millions and billions of teeny tiny pieces of matter called atoms (taken from Democritus's original word *atomos*). But that discovery wasn't made until 1803, when an English scientist named John Dalton figured it out.

Dalton's breakthrough is often referred to as *Atomic Theory.* What he proposed was that all the atoms in one element (say, carbon) are identical to each other and that all the atoms in another element (for example, hydrogen) are identical to each other too. But what Dalton couldn't figure out is why carbon atoms differed from hydrogen and vice versa.

Despite not knowing everything—yet—scientists at the time accepted atomic theory while simultaneously trying to disprove it. (Spoiler alert: they couldn't because Dalton is, and was, correct.) Over the next century, they performed experiment after experiment in an attempt to poke holes in Dalton's theory. But all the data continued to support Dalton's hypothesis for atoms in elements.

At one point, a trio of scientists named Joseph Louis Gay-Lussac, Amedeo Avogadro, and Jöns Jacob Berzelius embarked on a particularly painful endeavor when they tried to establish the atomic masses of each element—it was absolute chaos. With each scientist using different techniques and different standards, each of their published data sets were completely contradictory. It was such a mess that the scientific community elected to defer to Italian chemist Stanislao Cannizzaro to establish a universal standard for mass that was so desperately needed.

I'm totally biased, but if I had been an active scientist in the mid-1800s, I would not have spent a single second on that idea. As someone who loves to break things apart and put them back together again, I would have been investigating a much bigger question: If matter can be broken down into atoms, what's the atom made of? Still to this day, I'm not sure if scientists of the Victorian era were limited by their technology or simply had no interest in answering that question. Regardless, it wasn't until the late 1800s that Sir J. J. Thomson finally decided to look deeper into what made up an atom by experimenting with cathode rays.

To do this, he sealed a glass tube that contained two metal electrodes, which basically looked like a capped beer bottle with two long, thin chunks of metal inside. In his experiments, Thomson removed as much of the air as possible from his tube and then passed a voltage through the electrodes. When he did this, he could visibly see the electricity that passed from one metal to the other, which he referred to as a cathode ray.

Through these tests, Thomson was able to determine that the cathode ray was attracted to positive charges and repelled by negative charges. More importantly, by changing the types of metal he used for his electrodes, he learned that the cathode ray was always the same, regardless of the element.

Thomson was extremely happy with these results because he knew they indicated a groundbreaking discovery. If the cathode ray was not unique to each element or atom, then it must represent one of the building blocks used to *form* an atom—even atoms of different elements. However, knowing that his fellow scientist John Dalton had just convinced everybody that atoms were unique, Thomson worried—correctly—that people wouldn't believe him. So, he kept experimenting.

Through a variety of intense calculations, Thomson discovered that the cathode ray he was using was significantly lighter than the mass of any known atom. Just like if you were to measure the mass of all the doorknobs in your house, they would be much, much smaller than the total mass of your house. This would be true for your neighbor's house, your parents' house, etc. Thomson discovered that every "house" (i.e., atom) contains a bunch of doorknobs that are identical, and that are always lighter than the overall mass of the house.

For Thomson's experiment, this meant that he had isolated a very small piece inside the atom. In fact, he had just discovered the electron! These teeny, tiny particles are located inside the atom and they carry negative charges.

I'm going to jump ahead in scientific discovery and share that an atom is made up of three small pieces: electrons, protons, and neutrons. Protons (which carry positive charges) and neutrons (which are, you guessed it, neutral) are located inside the nucleus (the center of the atom), whereas the electrons exist outside of the nucleus. In other words, if my body is an atom, my liver and kidneys would be my protons and neutrons. My electrons would be everything on the outside of my body, such as my jacket and gloves.

Just like it would be very easy for me to give my jacket to someone or let them borrow my gloves, atoms can easily trade electrons. It would not, however, be easy for someone to take my liver or my kidneys. Is it possible? Yes. Would I be the same after the surgery? No. Similarly, it is *extremely* difficult to transfer protons.

The number of protons in the nucleus of an atom defines what element it is. For example, a carbon atom will always have six protons in the nucleus, and a nitrogen atom will always have seven protons. If a nitrogen atom somehow happens to lose a proton, it would no longer be nitrogen. That atom becomes carbon because carbon has six protons. This process—what's known as nuclear chemistry—does not happen very easily. In fact, most of the time an additional neutron has to be shot at the atom in order for it to experience nuclear decay. This methodology is currently being used to generate energy (i.e., electricity) in nuclear power plants.

While it's rare for atoms to add or lose protons, they *love* to exchange electrons, and a lot of that has to do with how atoms are structured.

Picture how you dress for a cold winter day out in the snow. If you're an atom, we already discussed how your liver and kidneys would be the nucleus, which is where your protons and neutrons are. But now let's take a deeper look at your layers of clothing. Your innermost layer—your thermal underwear—would be the first layer of electrons. Your shirt and pants would be the next layer of electrons, followed by your jacket and snow pants.

The electrons that hang out in the "jacket" layer—called the outermost shell (or outer shell for short)—are extremely important in chemistry. They are called the *valence* electrons, and they are the electrons that are easily exchanged with other atoms in chemical reactions. Just like layers keep our bodies protected from cold temperatures in the winter, the outer shell protects what's on the inside of an atom—what's called the inner shell—from external forces.

The electrons in the inner shell *cannot* react with other atoms because they are shielded by the valence electrons. Just like how your coworkers cannot see your underwear because it is "shielded" by your shirt or jacket.

This works out well for the atom because each layer of electrons is negatively charged, therefore they repel each other. This means that there will always be tiny gaps between each of the electron layers within the atom—just like how there is always a small gap between our shirt and jacket.

To take this metaphor even further, atoms have different sizes and that all comes down to how many layers an atom is "wearing." We all know someone who has to wear a ton of layers to stay warm in the cold, whereas someone else can walk around in shorts and sandals all year long. The same rule applies to atoms: smaller atoms don't wear very many layers at all, whereas big atoms have lots and lots of layers.

When I refer to *valence* electrons, just remember that these are the "jacket" electrons on the outer shell of an atom. And just like you would take off a jacket on a sunny day to get some warmth directly on your skin, these electrons are ready to leave their outer shell and react with external forces too.

This might sound shocking, but scientists didn't catch up to what I just explained until 1932. A lot of that is because scientists worked for centuries in isolation and with limited information (remember that this was pre-internet). Until very recently chemistry was a slow, monotonous process. But luckily, we now understand that atoms are made up of protons, neutrons, and electrons—and that electrons can easily be exchanged between atoms. Also, at that time, scientists around the world realized that there needed to be a uniform way to organize what everybody had learned about each type of atom.

And that's when the periodic table was created.

The periodic table is more than just a reference for your science class. For scientists like me, it's essential because at one quick glance, it tells me everything I need to know about an individual element, its characteristics and how that element's atoms are likely to behave and interact.

Let's start with the basics. When the periodic table was first being designed, each element needed to be assigned a chemical name and chemical symbol. This might seem pretty straightforward and simple, but it wasn't. What happened a lot was that two people discovered—or claimed to discover—the same element around the same time but gave it different names. The question then became: What's the official name? As you can imagine, a lot of fights were provoked when, for example, panchromium was named vanadium or wolfram was named tungsten.

As recently as 1997, the US, Russia, and Germany viciously fought over the names of elements 104 to 109. In 2002, the International Union for Pure and Applied Chemistry (IUPAC) finally put an end to these shenanigans and put out recommendations on how elements should be named in the future. These recommendations are followed religiously now, but it can still take up to ten years to get an official name for a new element.

Figuring out a chemical symbol for each element was much easier because it's an abbreviation of the name. Most are obvious, like H for hydrogen or C for carbon, but some are less predictable, like iron. Its chemical symbol is Fe—taken from the Latin word *ferrum*. Two other chemical symbols that might come up on trivia night are W for wolfram (tungsten) and Hg for hydrargyrum (mercury).

After each element is given a name and a symbol, it receives an atomic number. The atomic number aligns with the number of protons in the nucleus. Hydrogen has an atomic number of one, which means it only has one proton in the nucleus. The highest atomic number at this time is 118. This element is called oganesson (Og), and it has 118 protons in the nucleus.

This means that oganesson must also have 118 electrons *outside* of the nucleus. That's because the atomic number of an element also indicates how many electrons are outside of the nucleus. The important thing to remember is that all elements are assumed to be neutral. That means the number of protons on the inside of the nucleus is the same as the number of electrons on the outside. So, if we were to look at the atomic number for hydrogen—1—we know that it has one proton inside and one electron outside. To get a little more technical, the one proton on the inside has a positive (+1) charge, which cancels out the negative (–1) charge making the element neutral (0). You could do the same math for oganesson (118 + –118 = 0).

Unfortunately, neutrons aren't that simple. The number of neutrons varies from atom to atom, even atoms of the same element. Therefore, chemists decided to add yet another number to the periodic table. What's known as the atomic mass represents how many protons and neutrons are inside the nucleus of any given element. Unlike the atomic number, the atomic mass is rarely a whole number. That's because scientists use the *weighted average* of the number of neutrons in an atom and add it to the number of protons to determine the atomic mass.

In general, individual atoms maintain a proton to neutron ratio that is relatively close to 1:1. This means that we can estimate the atomic mass by doubling the atomic number. For example, magnesium has the atomic number 12 with an atomic mass of 24.31 (from 12 protons and a weighted average of 12.31 neutrons), and calcium has the atomic number 20 with an atomic mass of 40.08 (from 20 protons and an average of 20.08 neutrons).

But just like with all things in science, there are exceptions to every rule. For example, uranium has an atomic number of 92 so I would expect for it to have an atomic mass around 184. Instead, it has an atomic mass of 238.03 due the number of isotopes of uranium that contain varying numbers of neutrons. Most atoms, like uranium, have several *isotopes*, and isotopes occur when two or more atoms of the same element have a different number of neutrons. Since none of the isotopes are considered to be "better" than the others, we group all of the atoms together and simply average the number of neutrons. This average number is used in our standard atom notation.

In uranium's case, we refer to it as uranium-238. Magnesium and calcium are magnesium-24 and calcium-40, respectively.

ISOTOPES

I like to say that isotopes are atoms with personality. An isotope occurs when two or more atoms of the same element have a different number of neutrons. Isotopes are actually really common, but we tend not to focus on these as much when teaching chemistry because neutrons are neutral. Therefore, they don't really affect how an atom behaves in a typical chemical reaction. (Instead, we focus on what does: protons and electrons.)

That being said, scientists have characterized every isotope ever discovered, which I think is pretty cool. Like Lady Gaga, isotopes are "born that way" and exist naturally on earth with extra neutrons.

A great example of this in action is carbon. The majority of carbon atoms have six protons and six neutrons. However, *some* carbon atoms naturally contain seven or even eight neutrons in the nucleus. These extra neutrons do not necessarily make the carbon atoms more reactive or more stable, but they do make them all isotopes.

It's like how two dogs of the same breed can look the exact same, but one Dalmatian might have a few more spots than the other. The two dogs are nearly identical, and the "extra" spots do not change much about the dog or the breed. That's exactly the same as isotopes—extra neutrons don't usually change the atom or the element or even its reactivity with other elements. It's just an extra definition.

Once scientists compiled the chemical name, the chemical symbol, the atomic number, and the atomic mass for each element, they wanted to organize them in a way that would help them make predictions about chemical reactivity. They needed to know how each element would react in order to avoid dangerous reactions, like creating toxic gases or blowing themselves up. And the best way to do this was to identify commonalities between the atoms by grouping them by their physical and chemical properties.

Several attempts were made to arrange the elements in a logical order. A German chemist named Johann Döbereiner tried to arrange all the elements in groups of three, and quickly noticed that the bigger atoms were often more explosive. Shortly thereafter, another German chemist named Peter Kremers tried to conjoin two triads by forming a T shape. The problem with the triad method is that the scientists had to keep track of a plethora of perpendicularly shaped triads, and there was no easy way to compare one group to another group.

But it was two scientists working independently—Dmitri Mendeleev and Lothar Meyer—who realized that they could organize *all* of the atoms in *one* table if they just ranked them by increasing atomic mass. With this method, they put all of Kremers's different triad T shapes together—like a puzzle—to get the first table of elements.

The unique part about Mendeleev's rendition of the periodic table is that it included two "new" elements. While putting it together, Mendeleev noticed that there was a pattern among the atomic masses of the known elements, and realized that he needed to leave a gap in his table for two more elements that had yet to be discovered. For example, let's say your math teacher asked you to identify the missing number in the following pattern: 2, 4, 8, 10. Hopefully, you would see that the number 6 is missing from the pattern, and that it should be 2, 4, **6**, 8, 10.

Mendeleev basically did the same thing. He had groups of atoms that had the same number of valence electrons, but the pattern in their atomic masses was not quite right. Therefore, Mendeleev proposed that not only had we yet to discover certain elements, but he was also able to predict their relative atomic masses. And like a lot of the scientists I've mentioned so far, Mendeleev's hunch was right. When gallium (Ga) and germanium (Ge) were isolated and identified in 1875 and 1886, respectively, Mendeleev was finally given the long-overdue credit for creating the first true periodic table.

Today, the periodic table we use is based on what Mendeleev created. It contains seven rows and eighteen columns of little boxes. Each box represents an element and contains the same four standard pieces of information that scientists used to characterize elements way back when: the chemical symbol, the chemical name, the atomic number, and the atomic mass. Having all this information at our fingertips, chemists like me—and you—can determine the number of protons, electrons, and valence electrons that an atom has within a second.

The table is critical for scientists because it gives us an incredible amount of information about the elements that make up all matter in the world. It's *so* important that my university threw a party last year to celebrate its 150th anniversary. We had a periodic table made from cupcakes, I performed a few demos, and the dean of our college made a beautiful speech. It was one of the nerdiest parties I have ever attended, and to be honest, I loved every minute of it.

There's a periodic table at the back of this book, but if you want an electronic version, I highly recommend ptable.com. I am going to reference the table several times throughout this book, so I want to make sure *you* know how to use it too. This table is going to guide us through our section on health and wellness, and be crucial in our analysis of the chemistry we interact with in our everyday lives. We need to know the position of the elements on the periodic table, and what that means to us in terms of reactivity. Being able to understand the periodic table will help to understand why you should always use the same brand of shampoo and conditioner, or why your cakes don't look like what you see on *The Great British Baking Show*.

Let's look at an example. Please flip to the periodic table and find the box with hydrogen's chemical symbol, H, in the upper-left side of the table. If you look in the top corner of the H box, you will also see the number 1. This is the atomic number of the element, and it is always displayed in the top half of the box. You should also be able to find the number 1.008 in that same H box. This is the atomic mass of the atom, and it is always displayed in the bottom half.

You may notice that hydrogen is at the top of a bigger column. Each of the columns on the periodic table are called groups or families, and the column number indicates the number of valence electrons that each of the elements contains. (Remember, valence electrons are on the outer shell, like a jacket.)

HOW TO TALK LIKE A CHEMIST

If you want to sound like a chemist, drop the 10 in the column numbers of the periodic table. Most scientists refer to *groups* 3, 4, 5, 6, 7, and 8 instead of columns 13, 14, 15, 16, 17, and 18, respectively. That's because the group number represents the number of valence electrons. We do not do this for columns 3–12 because the elements in these groups do not always follow the traditional rules for valence electrons. But for columns 13–18, we use the easy shorthand notation because the number of valence electrons lets us predict how that atom will behave in different environments.

For example, hydrogen is in column one, therefore it can only have one valence electron. For this reason, lithium, sodium, and all the other elements in group one also must have one valence electron. That means we can expect for all group one elements to behave very similarly in equivalent environments. I can tell you that hydrogen (and all other elements in group one) likes to donate its electron to other atoms and that it is going to be extremely reactive. But why is that?

For a nonscientist, it would be logical to think that an element with one valence electron would do everything possible to protect (and keep) its only valence electron. However, that is actually the exact opposite of how most atoms operate. Instead, the electron is pushed away from the nucleus. Weird, right?

Let's break down this concept a little further. If we know the nucleus (your liver and kidneys) is positively charged, then your electrons (your shirt and jacket) will be greatly attracted to your positive core. However, when more electrons are added to the atom, there is a greater possibility of having electron-electron repulsions. In other words, your shirt will repel your jacket. So instead of the nucleus trying to desperately hold on to its one or two valence electrons, the inner shell pushes the valence electrons off the atom (your shirt pushes your jacket off your body).

For this reason, most elements with two electrons are also quite reactive. They are slightly more stable than elements with one electron, but in general, the elements in group two have no problem giving up their electrons. Beryllium, magnesium, calcium, and strontium are all great examples of elements with two valence electrons that experience the same electron-electron repulsions as group one elements.

Carbon and silicon are both in column four so they each have four valence electrons. That means we can expect for carbon and silicon to behave very similarly in equivalent environments.

Since chemists already know that carbon and silicon are quite stable, we would expect for any element in group four to be stable—just like we see with germanium, tin, and lead.

Mendeleev was prescient that chemists in the future would want to predict how elements would react with one another. That's why he organized the periodic table we still use today by valence electrons *and* atomic mass. (It's also why the periodic table is shaped like a bowl instead of a rectangle. The big gaps at the top of the periodic table allows for the elements to be arranged by their physical and chemical properties).

As you move down within any column on the periodic table, the atoms get larger and larger. Generally speaking, the largest atoms are in the bottom left corner of the periodic table and the smallest atoms are in the upper-right corner.

Each of the rows—or *periods* (hence the name periodic table)—on the table represent an extra "layer" of electrons for that atom. As you move across a period on the periodic table (from left to right), the atoms typically get smaller and smaller. Seems backward, right? How can it be possible that helium is smaller than hydrogen?

As you move across a period, every element gains one additional proton and electron. This means that the positive charge in the nucleus increases every time the atomic number increases. The greater the positive charge, the more attracted the valence electrons are to the center of the atom (i.e., the nucleus).

For example, hydrogen has a nuclear charge of +1. Since it is in group one, we would expect for hydrogen to have one valence electron as well. This means that the +1 charge in the nucleus is attracted to the –1 charge on the electron.

But now let's compare that to the attractions within a helium atom. Since helium is in group two, we would expect for it to have two protons and two electrons. This attraction between the +2 charge in the nucleus to the –2 charge from the valence electrons is significantly larger than the attraction between hydrogen's +1 and –1. This means that helium's valence electrons are sucked toward the nucleus more so than hydrogen's valence electrons will be. Therefore, helium has a smaller atomic radius than hydrogen.

If we combine the repulsions between electrons and the attractions between protons and electrons, we can begin to pick up on a few periodic trends. An easy way to remember how these groups and periods work is that "francium's fat." As one of the largest atoms on the periodic table, francium is located in the bottom left corner with the atomic number 87. It has 87 protons, 87 electrons, and an average of 136 neutrons. If francium was a person, she would be wearing a LOT of clothing.

There's one more thing you can figure out *just* by looking at the periodic table: how receptive an atom is to change. Remember, atoms can lose or gain electrons pretty easily—it's like taking off a jacket, or for bigger atoms like francium, a layer of clothing.

We describe an element's willingness to gain or lose an electron as its *electron affinity*. For example, most of the elements in the upper-right corner, like oxygen and fluorine, have large electron affinities, meaning that they desperately want to gain an electron. The elements in group seven (column seventeen) are notorious for searching for an electron to steal from a neighboring atom, with fluorine being the most reactive.

WHAT'S *AN*-ION ANYWAY?

When an atom has gained (or lost) an electron, we call it an *ion*. We use the term *anion* to represent any atom that has gained one or more electrons and the term *cation* to represent any atom that has lost one or more electrons.

Let's look at anions first. An anion is always negative and it always has more electrons than protons. It will also be bigger than the corresponding neutral atom. If my husband were to give me his big, puffy coat to wear, I would appear bigger. Similarly, an atom that has gained an electron (now called an anion) becomes bigger. A great example of this is fluorine. Fluorine atoms always want to gain one electron to convert into the fluoride anion (F^-). When fluorine is neutral, it is useless to the human body. But once fluorine gains an electron and becomes fluoride (an anion), it becomes a micronutrient that can help prevent cavities by encouraging healthy bone growth in the human body. It's fascinating to me that one tiny electron can make huge differences in the chemical properties of an atom.

The term *cation* is used to classify any atom that has lost one or more electrons. In the puffy coat example, my husband represents a cation by giving me his jacket—an electron. Cations are always positive and have more protons than electrons. Cations also appear to be smaller than the original neutral atom—just like my husband would appear to be smaller after giving me his jacket.

Unlike common anions, the atoms most likely to become cations are located in the upper-left corner of the periodic table, like lithium and beryllium. These elements have one or two valence electrons that can easily be donated to another atom. That's why these elements are much more likely to become cations than anions.

This is particularly true for the elements located in group one, especially lithium. The lithium atom only needs to lose one electron to convert to the lithium cation (Li^+). In the form of the ion, lithium cations can be used to treat bipolar disorder by helping to manage the brain's sensitivity to dopamine, whereas the neutral lithium metal has no advantages in the human body. Once again, we see that adding or subtracting just one electron can drastically change the physical properties of the atom.

One last category you should know about is group eight (column eighteen). All of these elements are what we call inert, or inactive. These elements do not want to gain or lose an electron. I think of elements in this group, like helium and neon, as the people who like to chill at home on a Saturday night versus go out to a party. All elements in group eight (helium, neon, argon, krypton, xenon, and radon) are actually called the noble gases because they so rarely interact with the other elements—like royalty.

The periodic table does so much more than just give us a cheat sheet. By looking at it, we're looking at centuries of discoveries by thousands—if not hundreds of thousands—of scientists all around the world. Using it, we can do amazing things, like create imaging that detects cancer and invent semiconductors that work in solar panels. Even the lithium-ion batteries in your cell phone and laptop are a result of the patterns on the periodic table—they only function because of the electrons moving within (and between) atoms. In fact, with a solid foundation of the structure of an atom, it's easier to see how these electron-proton interactions show up in the world over and over again.

Now that you understand the basics of an atom—its protons, neutrons, and electrons—as well as how atoms make up an element, we can move on to what happens when two atoms of different elements come together. This is where chemistry starts to get really exciting, because the attraction between atoms is a lot like dating or making a new friend.

Will there be an attraction?

How will the two react?

Can they form a bond?

2

ALL ABOUT THE SHAPE

Atoms in Space

In the previous chapter, you learned how atoms are the building blocks of basically everything in the universe. But how do those building blocks come together to form, say, a computer? Or salad dressing? Or an ice cold beer?

Electrons.

When two or more atoms join together, it's by sharing or transferring electrons through a bond. And anything with a bond is a molecule or a compound. A single atom is never a molecule or a compound, it's always just an "atom."

And before we jump into chemical reactions, you need to know that chemists refer to a collection of molecules as species, substances, and even sometimes, the system. These terms are interchangeable and all mean the same thing—that we're talking about a collection of molecules. So, when I talk about a *species*, you'll know I'm talking about a group of molecules, whereas when I refer to a *molecule*, you know it's all by itself.

Cool?

Cool.

We can see atoms form bonds every day, if we know what to look for—like when salt dissolves in the ocean or how a face mask removes blackheads. Atoms bond with one another based on attractions. In that way, atoms are just like us! Because protons have a positive charge and electrons have a negative charge, their bond neutralizes both atoms, which is exactly what atoms want.

Atoms feel an attraction toward one another when they are physically close to each other. Since electrons are on the outside of the atom and protons are on the inside, there are actually two attractions happening at one time.

Let's pretend we have two atoms: A and B. The electrons in atom A will be attracted to the protons in atom B, and the electrons in atom B will be attracted to the protons in atom A. Generally, the only thing that can get in the way of this is electrons repelling other electrons.

Atoms can muck up a potential bond by getting too close, just like a stranger could turn us off by sitting too close in a coffee shop. When someone we don't know invades our personal space, we usually establish more distance to feel comfortable again. Sometimes that means getting up and walking away, and that's exactly what happens with atoms. If the electrons of one atom get too close to the electrons of another, the electrons repel and make more space.

Eventually, two atoms will establish the perfect distance, where the attraction between protons and electrons overpowers the repulsion between electrons. In other words, the proton-electron attractions are maximized and the electron-electron repulsions are minimized. When this happens, a bond can form.

Let's assume that you and the stranger in the coffee shop find a comfortable distance too and start chatting. If you're attracted to one another, the next logical step would be to form a more permanent connection. Now, in real life you'd probably just have another cup of coffee or get the person's phone number. But, since this is an analogy of what happens when atoms bond, we'll pretend that the next step is to hold hands.

When atoms "hold hands," they are actually forming a bond. In chemistry, a bond is essentially an agreement between two atoms. The atoms will go everywhere together until a more attractive atom comes along. For example, if I am holding hands with the cute stranger, I will continue to do so—until Ryan Reynolds walks into the room. At this point, I will then drop the cute stranger's hand and be off to pursue a better connection. That *also* happens with atoms.

But here's the difference. I could walk off into the sunset with Ryan Reynolds and be the exact same Kate that walked in the door of that coffee shop *and* the same person who held hands with the stranger. Neither Ryan nor the stranger took my arm or leg, right? Unfortunately, that is not always true for atom A and atom B.

Unlike with me and the stranger, when two atoms decide to bond, the individual atoms are no longer viewed as independent entities. Instead, when atoms form a bond, there is an instant exchange of electrons. So sometimes after atom A and atom B break apart, atom A can carry around an electron or two from atom B.

But when they stay together, we try to analyze how equally the electrons are shared between two atoms within the bond. And to do that, we have to look at the *character* of the atom by examining the atom's composition. The easiest way to do that is to figure out whether the atom is classified as a metal or a nonmetal. Luckily, it's usually pretty easy to tell the difference between these two types of elements just by looking at them—in a lab or real life.

Most metals are very pretty, especially after they have been properly cleaned. Elements that are defined as metals like gold, cobalt, and platinum are shiny and lustrous, because most reflect light very easily. Many metals are bendable and malleable, which makes these elements perfect for crafting jewelry. (We use the term malleable to describe a metal that can be hammered into another shape.) Metals also conduct heat well, which you've probably learned if you've touched a hot metal pan on the stove.

This group of elements is also known for being wonderful electrical conductors, which means that electrons can easily move through most metals very quickly and with little resistance. That's why it's a bad idea to stand around with an umbrella in a thunderstorm. The metal typically used in the handle (and on the top) attracts the electricity associated with lightning. And because metals are good conductors for electricity, these electrons are actually what electrocute people. On the other hand, we take advantage of this property all the time, like when we're using the battery in our smartphones.

Metals like to donate their electrons to other atoms, but do not like to form bonds that force them to gain electrons. Metals are like Santa—they love to give things away but hate to receive. (Unfortunately, there's no equivalent of milk and cookies for atoms.) Additionally, because metals would have to gain an electron in order to bond with another metal, these elements usually avoid that circumstance too.

In contrast, nonmetals are not shiny, not malleable, and not ductile. The term *ductile* is used when a substance (usually a metal) can be pulled into thin wires. What defines nonmetal elements is the fact that they're not metal. (Obvious, I know.) The majority of solid nonmetals are dull and boring. Gaseous nonmetals are mostly colorless, which means that we can't even see these elements, much less make pretty jewelry from them.

What you need to know about nonmetals is that electrons don't move through these substances very easily. Nonmetals are bad at conducting heat and electricity. It's hard for the electrons in a nonmetal to move, which is why a lot of them are unreactive. (This is also why all the noble gases you learned about in the last chapter tend to hang by themselves.) Simply put, their electrons cannot pass from atom to atom as easily as they do between metals.

You can find most nonmetals in the upper-right corner of the periodic table, starting with carbon in group four. Nonmetals go all the way to group eight. For each of the periods below carbon, the other nonmetals are located to the right of silicon, arsenic, tellurium, and astatine.

There are more than five times the number of metals than nonmetals, but 99% of the universe is made of hydrogen and helium—two nonmetals! Another nonmetal—oxygen gas—is crucial for human survival. The most fascinating part about nonmetals is that some of them are extremely stable, while others are unbelievably reactive.

The reason I'm talking so much about metals and nonmetals is because the composition of the atom (is it a metal? Or not?) is the first question we ask when trying to figure out what type of bond it will have within a molecule. There are two main types of bonds in chemistry: covalent and ionic.

Let's start with covalent bonds.

The simplest form of the covalent bond is what's called a single bond.

A single bond is formed when two atoms share two electrons. In fact, all covalent bonds are formed when two atoms share electrons. For a single bond, each atom usually donates one electron. Let's go back to our previous example, and look at the bond that I formed with Ryan Reynolds.

To demonstrate a single bond, let's pretend that Ryan uses his left hand to hold my right hand. We have two electrons between us and are at an arm's length from each other. At this distance, I can begin to feel the pull of my "electrons" toward his "protons."

Now to form a double bond, Ryan takes his empty right hand to grab my left hand. When he does this, I have to turn my body to hold his other hand. This movement decreases the distance between Ryan and me—because now, we're standing face-to-face. Our "connection" is now twice as strong because we have two bonds between us. (Hence the name double bond.)

The double bond is much stronger than the single bond, and because of the way the electrons are arranged, the atoms can tolerate being slightly closer together. Within a double bond, there are four electrons between the two atoms—one in each pair of hands being held.

For a triple bond, Ryan would need to wrap his leg around my body (please don't tell my husband). A triple bond allows for atoms to get ridiculously close together. Ryan and I now have three bonds between us—two within each pair of hands and where his leg is wrapped around my body. That gives us three places to share electrons.

If we do the math, we know that three bonds with two electrons each creates a total of six shared electrons between the two atoms. That's part of the reason why a triple bond is very strong and very difficult to break—much more so than a single or double bond. Additionally, the distance in a triple bond is much smaller between the atoms, because they're sharing six electrons.

The single, double, and triple bonds are the most common types of bonds in covalent molecules. You interact with these bonds all the time. In your shampoo and your toothpaste and your morning coffee—even in your clothing, your makeup, and your deodorant. As I'll explain later in the book, covalent bonds are everywhere in your life, wherever you happen to be.

If you look up right now, most things in your vicinity contain a covalent bond. And I don't even know where you are! That's how ubiquitous covalent bonds are in our world.

Scientists evaluate covalent bonds by looking at how the atoms actually share electrons. Is the sharing even? Or does one atom tend to hog all of the electrons? When two atoms share the electrons perfectly equally, the bond is referred to as a pure covalent bond. This can only happen when the electrons on atom A are as attracted to the protons in atom B as atom's B protons are attracted to atom A's electrons. That's a mouthful, right?

It might be easier to think about pure covalent bonds like we do a romantic relationship.

I can form a pure covalent bond with someone if my heart is as attracted to his body as *his* heart is attracted to *my* body. How badly are his insides attracted to my outsides?

If the attraction is equal, then a pure covalent bond is formed.

Just like in love, it is very rare that the attraction is perfectly equal between two atoms. Instead, most attractions are slightly unbalanced. When the attraction between two atoms isn't equal, that's no longer a pure covalent bond. Instead, that's classified as a polar covalent bond. Now we're getting into the electricity of attraction—and I'm not talking about the spark you feel when you meet someone super cute. Electronegativity is how chemists quantify *how* attracted atom A's electrons are to atom B's protons. Polar covalent bonds are produced when two atoms have different electronegativities, while *pure* covalent bonds have atoms with the *same* electronegativity.

Still with me? To recap, in a pure covalent bond both atoms are equally into one another. But in a polar covalent bond, one of the atoms is more attracted—or more electronegative—than the other. Scientists generally know how electronegative an atom is thanks (again) to the periodic table. Electronegative atoms are located in the upper-right hand corner, with fluorine, oxygen, nitrogen, and chlorine. These four atoms are attractive to *a lot* of other atoms. In contrast, the most electropositive atoms—which don't attract many atoms—are in the upper-left corner of the periodic table. Lithium, beryllium, sodium, and magnesium are all electropositive atoms.

Chemists want to know which atom is stronger (or more electronegative) within a polar covalent bond because we always want to know where the electrons are hanging out. The position of these electrons within a bond determines how that molecule will interact with another molecule. Remember, chemists are *obsessed* with predicting the outcomes of chemical reactions.

A lot of scientists find molecules with an even distribution of electrons kind of boring, because these species tend to be unreactive and only hang out with other evenly distributed molecules.

Molecules with an uneven distribution of electrons, however, tend to be extremely reactive. These species are cool to chemists like me because they prefer to interact with other reactive molecules.

For now, let's pretend that the periodic table shows that Ryan Reynolds is the less attractive (electropositive) partner in my bond with him. Once we know that I am more electronegative than him, we can also predict that his valence electrons will attempt to leave his body and travel instead toward mine. The electrons would travel from his arm, through our covalent bond in our hands, and continue up my arm until they are sitting on top of my shoulders. The electrons would then remain on my body until our bond is broken. At that point, the electrons could choose to either jump back to his body or leave with me forever.

Let's look at this interaction in real life. When a bond forms between carbon and fluorine (C–F), scientists would first look to the periodic table in order to confirm which atom is more electronegative. (In this case, it's fluorine). That tells us that carbon's valence electrons are likely to leave carbon and move as closely to fluorine as possible through their covalent bond.

Since electronegative atoms carry the majority of the electrons in the bond, they are often given a partially negative symbol (δ–). The electronegative atom in the bond attracts electrons, therefore it carries a partially negative charge. This means that the electropositive atom—or the less attractive atom which has just lost some electrons—will be partially positive (δ+). The word *partially* indicates that electrons are still being shared between atoms—usually in the covalent bond (the atoms' "hands").

This is a direct contrast to the bonds formed between a metal and a nonmetal. Just like covalent bonds, the metal-nonmetal bonds are formed when atoms get close enough to be attracted to one another. But unlike covalent bonds, this new type of bond is only formed when the electrons have been *transferred* from one atom to another. More specifically, when a metal transfers an electron to a nonmetal. And when this happens, an *ionic* bond is formed.

It is very important to understand that, unlike in a covalent bond, ionic compounds do not share electrons. These atoms transfer the electrons, which results in the formation of a positive metal ion and a negative nonmetal ion (unlike the partial charges found on atoms in covalent bonds). And remember, opposites attract, so the metal cation is now greatly attracted to the nonmetal anion.

If a covalent bond is two humans attracted to each other in a committed relationship where love flows back and forth, then an ionic bond would be the kind of relationship where one partner is always giving and the other is always taking. An ionic bond is very one-sided where the cation (with less electrons) is the giver and the anion (with more) is the taker.

Like covalent bonds, ionic bonds are everywhere in the world around us. For instance, table salt is formed by the ionic bond of a sodium atom and a chlorine atom. When sodium (a metal) gives its electron to chlorine (a nonmetal), the sodium atom becomes a cation and the chlorine atom becomes an anion. In table salt, the chlorine is the partner that's taking and the sodium is giving.

Now that you know the basics of how atoms bond together—either in covalent or ionic bonds—we can move on to something else that's super cool about molecules.

THE SECRET FORMULA

In chemistry, we use molecular formulas to represent the atoms in a molecule. There are two types of formulas: condensed and structural. Most people are familiar with the condensed molecular formula, which tell us exactly which atoms are in the molecule and in which ratio to one another.

Let's talk about H_2O, shall we? Water has two hydrogen atoms and one oxygen atom, which is why its condensed molecular formula is H_2O. The subscripted 2 is listed after hydrogen because there are two hydrogen atoms in water. In condensed molecular formulas, the subscript is always listed *after* the atom it references, which tells us the number of each atom in the molecule.

However, condensed formulas do not give us any information about the bonding within the molecule. If you were to look at the molecular formula H_2O, you may (incorrectly) assume that the molecule looks like this: H–H–O. This formula suggests that the two hydrogen atoms are bonded to each other, but in reality, water is formed when each hydrogen atom bonds directly to one oxygen atom, like this: H–O–H. There is no way to look at H_2O and know how the hydrogen and oxygen atoms are bonded (unless you have a strong background in chemistry).

Instead, we use the other kind of formula—structural molecular formulas—to indicate the arrangement of the atoms. Since each hydrogen atom is bonded to the central oxygen atom, its *structural* formula is HOH. This type of formula indicates that hydrogen A is bonded to the oxygen atom, which is bonded to hydrogen B, like this: H–O–H.

But how do you know which formula to use?

Honestly, it really just depends on the context.

Structural formulas give us the most information, which is why it's the preferred molecular formula among chemists. However, for a molecule with lots of atoms, it is impractical to provide a structural molecular formula because it would be so long and cumbersome. Therefore, the most common way to report a molecule is to provide the condensed molecular formula.

Remember how I mentioned that double and triple bonds require a small distance between the atoms? That's because molecules have unique shapes. It might surprise you to learn that the shape of any given molecule isn't determined by the atoms it's made of. Instead, these shapes are based on what chemists are obsessed with.

Electrons.

Back in the 1950s, two chemists named Ronald Gillespie and Ronald Sydney Nyholm began to notice patterns in the shapes of molecules. Not surprisingly, they quickly learned that the geometry of the molecule was determined by the arrangement of the electrons in space, *not* the identity of the atoms. In 1957, Gillespie and Nyholm published what's called the VSEPR theory (valence shell electron pair repulsion), which accurately predicted the three-dimensional shape of any molecule based on the number and relative position of the electrons.

For example, we know that a molecule with two atoms will always take a linear shape. There is simply no other way to connect two atoms with one bond. All molecules containing just two atoms will be linear, regardless of what the atoms are made of.

Carbon monoxide is a classic example of a two-atomed molecule. The carbon and oxygen have a triple bond between the atoms, and since there are only two atoms, it always maintains a linear shape. This odorless, colorless gas is quite flammable, but it is also very dangerous. When you inhale it, the tiny molecule binds to the hemoglobin in your blood and kicks out any oxygen molecules. This is why too much of "the silent killer" can be fatal.

Through Gillespie and Nyholm's extensive research, this dynamic duo was able to extend this model to molecules with any number of atoms. The underlying idea that makes this theory work is something you already learned—that electrons will always repel other electrons.

I like to think of electrons as needing elbow room in the molecule, meaning each bond wants to be as far away from all of the other bonds in the molecule as possible. The position of the electrons in any given molecule is referred to as the electronic geometry of the molecule. And remember this is all about electrons—so the geometry is about how many electrons there are and where they're positioned within their bonds.

Gillespie and Nyholm proposed five main electronic geometries to describe the distributions of the electrons across a molecule. I know it may seem like the shape of the molecule is not that important, but it actually helps us to determine *how* the electrons are arranged within a molecule. Are they distributed evenly? Unevenly? When we combine the electronegativity factor with the overall shape of the molecule, we will finally be able to determine *how* two molecules react with each other.

FORMULA	SHAPE	STRUCTURE
AX_2	Linear	X—A—X
AX_3	Trigonal Planar	
AX_4	Tetrahedral	behind the page; in front of page
AX_5	Trigonal Bipyramidal	behind the page; in front of page
AX_6	Octahedral	behind the page; in front of page

Let's assume our molecules have one central atom (A) and any number of terminal atoms (X) bonded directly to A. For this discussion, the central atom will always be in the middle of the molecule, and the terminal atoms will always surround it. This means that a molecule with three atoms will have the molecular formula AX_2, with one A atom in the middle and two X atoms on the outside of the molecule.

According to VSEPR, the two X atoms in the molecule will try to spread as far apart as possible around atom A. One X atom will be on the left and one X atom will be on the right with a 180° angle between the bonds. Carbon dioxide is a great example of this shape, which is defined as linear. It is also the molecule that makes up dry ice, which is one of my favorite cryogenics.

Following the same rules, a molecule with four atoms would have the molecular formula AX_3, where all three X atoms are aligned perfectly around the central A atom. The geometry of this molecule is called trigonal planar because there's a 120° angle between each of the bonds. The word *planar* was added to the name to indicate that these molecules are flat like a piece of paper.

Formaldehyde (CH_2O) is one of the best examples of a trigonal planar molecule, and one of the most misunderstood chemicals. Not only is it made naturally in your body, but it is also commonly found in broccoli, spinach, carrots, apples, and bananas, which are all foods that are very good for you. However, formaldehyde is bad for you in high doses for long periods of time, therefore certain industrial workers are at higher risk of having adverse health effects.

These planar molecules are a direct contrast to the funky shapes formed by molecules with five atoms, even though we're still following the same rules. The shape of AX_4 is tetrahedral—with four faces. The X atoms are arranged to have the maximum distance between the nearest atom, resulting in bond angles of 109.5°. That's impossible to draw on a piece of paper because the tetrahedral shape isn't planar (or two-dimensional). Two X atoms can and will stay on the paper but one X atom would be raised above the paper, and one X atom pushed below the paper in order for the molecule to stay consistent with the rules. Remember VSEPR demands that atoms achieve the biggest distance possible between the bonds in space.

In other words, atoms in bigger molecules have to break the plane in order to prevent the electrons within the molecule from repelling one another. Methane (CH_4) is a classic example of a tetrahedral molecule. It's the gas that comes out of gas stovetops, but it is not the gas that you smell if you have a gas leak. (That gas is called methanethiol and it smells like rotten eggs. We began adding this completely harmless molecule to natural gas back in 1937 after the London School in New London, Texas, exploded from a gas leak, killing almost 300 students and teachers. The smell of methanethiol is so pungent that it gets people's attention *quickly*.)

Six-atomed molecules with the molecular formula AX_5 take a shape called trigonal bipyramidal. This complicated geometry gives us a molecule with one atom above the plane and one atom below the plane. Then imagine adding three more atoms spread 120° apart along the plane. Did I lose you? Let me try to explain this bizarre shape using the human body.

If your body was a trigonal bipyramidal molecule, the A atom would be your torso. There would be an X atom on your head and an X atom on your feet. Then you would have an X atom directly in front of your hips. You would also have one X atom shooting out of your left ass cheek and one X atom shooting out of your right ass cheek. It's an intricate molecule with a lot of unexpected symmetry.

Molecules with seven atoms have a very similar shape to molecules with six atoms. We still have one atom above the plane and one atom below the plane. But now, there are four atoms spread 90° apart along the plane—or four X atoms shooting from your left hip, right hip, left ass, and right ass. This shape is called octahedral because all of these molecules have eight faces.

The best example of an octahedral molecule—by far—is sulfur hexafluoride (SF_6). If you inhale this gas, your voice will drop to a much lower register to give you the opposite effect of inhaling helium gas. (It's also the gas behind "Fartgate" on *The Wendy Williams Show*. Look it up. The viral video involves yours truly).

VSEPR helps scientists to know how the electrons are arranged around the central atom of a molecule. But some molecules—like the caffeine in coffee, the ethanol in beer, and the carbohydrates in chips—have more than one central atom. In these cases, we combine the geometries of all the internal central atoms to determine the overall shape of the big molecule.

Let's look at an example of how this is done with molecules that contain more than fifty atoms, like cis and trans fats.

A few years ago, the United States Food and Drug Admin-

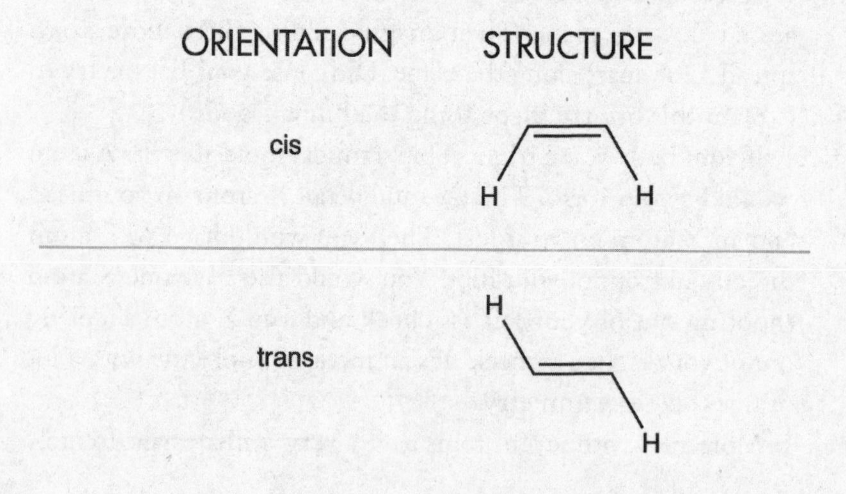

istration gave all food suppliers three years to find a way to remove trans fats from their products. As of June 2018, all trans fats have officially been banned from food in the United States. However, there are no restrictions on what are called cis fats. This can be surprising to some people because cis and trans fats have the same molecular formulas, and these two molecules can be made through very similar processes.

The *only* difference is the shape of the molecule. Trans fats are long and tubular (like a toothpick), while cis fats have a kink (like a toothpick that has been snapped in half).

When the trans fats get inside of your arteries, they can line up perfectly next to other trans fats. They begin stacking on top of each other and slowly begin to clog your arteries. Sometimes, they align themselves so neatly together, that they stop all oxygenated blood flow away from your heart. This can lead to a heart attack, among other adverse health effects.

You can visualize this by thinking about how easy it would be to stack a bunch of toothpicks together and then place them into one end of a hose. If they are tightly packed together enough, the water will not be able to make its way through the toothpick blockage.

But now think about what would happen if you take all of those toothpicks and snap them in half? Would you be able to stack all of them together again as neatly? Unlikely. No matter how hard you try, the snapped toothpicks will not be able to clog the hose as easily as the intact toothpicks did—just like how the cis fats are not able to clog arteries as easily as trans fats can.

Hopefully, with this example you can see that the shape of a molecule really matters in chemistry (and in your arteries). It tells us where the electrons are and how the molecule will operate in a 3D space. But more importantly, when we know the location of the electrons, we can begin to analyze how the electrons actually form the bonds between atoms within the molecule.

But to do this, we need to look at atoms a little more closely.

First, let's start by thinking about each layer of the atom as having pockets—pockets in your underwear, pockets in your shirt, and pockets in your jacket. Each of these tiny pockets represent what's called an atomic orbital. And each atomic orbital can hold two electrons at one time, maximum. These pockets can never, ever have three or more electrons. That's because there isn't enough space and because the pocket can't handle the charge of the third electron.

Remember, electrons repel other electrons and need their space.

In fact, even when there are just two electrons next to each other in one pocket, or orbital, there is a discomfort felt by both electrons. To minimize the repulsions felt by each electron, the electrons begin to spin in opposite directions: one electron spins clockwise and the other electron spins counterclockwise.

Try doing this right now with your hands. Have your left hand spin clockwise and your right hand spin counterclockwise. I do this for my students every semester, and I look silly trying to get my hands to move in opposite directions. My students always laugh at me, but here's the thing. I know it doesn't sound like much, but when the electrons spin in opposite directions, they stabilize the atom. Surprisingly, the spinning motion allows for the electrons to spread out as much as possible within the small orbital. In other words, the electrons are able to achieve the maximum distance between the two negative charges.

But at this point, my guess is you are thinking—so what? Why should I care about orbitals (and their occupancy rules)? How do atomic orbitals actually affect me in my daily life?

And to be honest with you, I get why you are asking that question.

The real-world applications of atoms and molecules are relatively straightforward. Just look at something as simple as your clothing. The molecules in the dyes give your shirt its red or blue color. The distances between the molecules dictate how easily a fabric breathes or removes sweat if you are wearing wicking materials.

But orbitals? Their science is more intricate, and in my opinion, more beautiful.

On the Fourth of July, we see electrons moving from orbital to orbital in every firework. The red fireworks are a result of small movements of electrons between orbitals, while green ones are from much bigger movements.

On Halloween, we see orbitals in action every time we observe phosphorescence—the chemical phenomenon that makes things glow in the dark. Whether we realize it or not, we are constantly observing electrons moving within their orbitals, and also between orbitals. We are just lucky that scientists have engineered ways for us to play with these movements safely—like sparklers and glow sticks.

All of this chemistry stems from four types of atomic orbitals, or pockets, on an atom where electrons can hang out. There are *s* orbitals, *p* orbitals, *d* orbitals, and *f* orbitals, and one scientist named Erwin Schrödinger proposed all of the atomic orbitals at one time, which is just mind-boggling. In one quick paper, he established so much about how atoms bond. In fact, not much has changed in the last hundred years. Chemists like me still operate under the assumption that there are four main types of atomic orbitals.

But remember—no matter how big or what shape the orbital is, it can only accommodate two electrons. And those electrons have to hang out at the biggest distance apart from one another (because of those electron-electron repulsions).

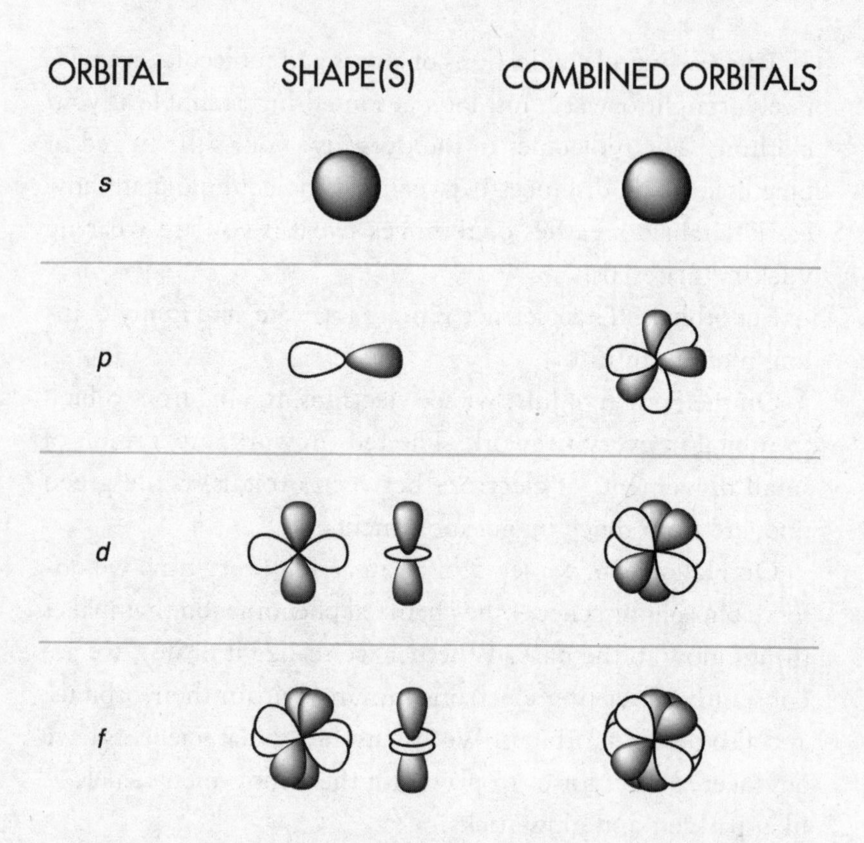

Electrons have the most freedom to move about in the *s* orbital because it is shaped like a big, round ball. It's just a simple sphere that perfectly surrounds the nucleus in the atom. While it may seem counterintuitive, the *s* stands for *sharp* due to the sharp peaks that are produced in the lab by *s* orbitals.

To look at a simple example, we can examine the lowest energy orbital inside an atom, which is called the *1s* orbital. Every single atom on the periodic table contains a *1s* orbital. It is the orbital closest to the nucleus and, as I mentioned before, can only contain two electrons. Since hydrogen and helium only have one or two electrons, respectively, all of their other atomic orbitals are completely empty. This makes hydrogen and helium ideal for illustrating why orbitals are so important.

Let's look at helium first. It has two electrons in its *1s* orbital and is considered to be a very stable element, which you may remember from our discussion on noble gases. Helium is so stable that we commonly use it for birthday balloons, hot air balloons, etc. There are no safety concerns about the element because it is extremely inert. This means that even if the wind catches the helium balloon and knocks it into the birthday candles, nothing dangerous will happen. The balloon will just pop and the helium gas will float up into the atmosphere.

But now let's look at hydrogen, an atom with only one electron in the *1s* orbital. The hydrogen atom is not stable at all, not even a little bit. This "open" space in the orbital makes monoatomic hydrogen an extremely dangerous atom. It is constantly searching for another electron to put into its empty *1s* orbital or for a way to give up its only electron. It is so reactive that monoatomic hydrogen is rarely found in nature by itself. Instead, it partners with another hydrogen to form diatomic hydrogen (H_2). Had we mistakenly filled our birthday balloon with hydrogen gas instead of helium, the flame from the birthday candle would have created a giant fireball instead of just popping the balloon. Whoops. Now *that* would be a party.

All due to one empty space in the atomic orbital—one gap in the atom's pocket.

As you might expect by now, similar reactions can happen when electrons are added or removed from the next atomic orbital: the *p* orbital. In this case, *p* stands for *principal*. The orbital is shaped like a figure eight, and it is often described as having two lobes. This just means that there are two sections where the electrons could be located in the *p* orbital. There are actually three identical versions of the *p* orbital on any given layer of the atom, clumped together to form a six-pointed star around the nucleus.

Each *p* orbital has a different orientation in space. The p_x orbital allows for the electrons to move left to right across the atom, the p_y orbital allows for the electrons to move forward and backward across the atom, and finally the p_z orbital allows for the electrons to move up and down the atom.

However, there is something, dare I say, *magical* about how the electrons move about the atom. Electrons never, ever exist in the nucleus, but are able to jump across from the left to the right side of the atom. Then the electrons move forward and backward, all without ever moving through the nucleus.

How do the electrons teleport from the left to the right side of the atom without ever transmitting through the nucleus? Honestly, we still do not know the answer to this question. There's still so much about chemistry that we don't understand, and this is one of those things that we haven't been able to figure out. I just hope that I'm alive when scientists learn how that part works.

When three *p* orbitals overlap, they form a starlike orientation that allows for six electrons (3 orbitals × 2 electrons per orbital = 6 electrons) to move about the atom with the maximum electron-proton attractions and the minimum electron-electron repulsions. If you look at the image of the *p* orbitals shaped like a six-pointed star, you will see that unlike the spherical *s* orbital, there are major gaps in this shape where the electrons *cannot* exist. The electrons have much more space—or freedom—to move about the *s* orbital than they do in the *p* orbitals. This is awesome for the electrons.

The next atomic orbital is my personal favorite, the *d* orbital. These orbitals are the basis for most inorganic chemistry. Each of the *d* orbitals have four lobes or four different spots for the electrons to exist. These orbitals kind of look like little flowers, where the nucleus is in the center, and the electrons are in the flower petals.

There are five different *d* orbitals, and four of them maintain that pretty flower shape. The only difference between these four flower orbitals is their orientation in space. To better understand this, let's look at the four-lobed *d* orbital.

If you put this book on a table, the *d* orbital is on a flat, horizontal surface (orientation 1). But now what if you stand up? You could put the book on the wall in front of you (orientation 2) or the wall to your left (orientation 3). Maybe you even used a divider that splits the room diagonally (orientation 4). Notice that the placement of the book would be in four different orientations in space: (1) flat, (2) vertical, (3) vertical but rotated 90°, and (4) vertical but rotated 45°. Each different placement of the book represents a different way that the *d* orbitals could exist on an atom.

The fifth kind of *d* orbital has a super weird shape that my former professor used to describe as a "sausage in a donut." Although it's an odd description, I have to give him credit because it's a perfect visual of the unique *d* orbital. Personally, I think it just looks like the p_z orbital wearing an inner tube around its waist.

All five of these *d* orbitals overlap to form a very intricate flower, just like the *p* orbitals formed a six-pointed star. However, this flower created from the *d* orbitals is a much more complicated network for electrons to move around in. The unique shape of the *d* orbitals allows for ten electrons (5 orbitals × 2 electrons per orbital = 10 electrons) to move about the atom with the maximum electron-proton attractions and the minimum electron-electron repulsions.

The last kind of orbital found inside an atom is called the *f* orbital, and it is by far the most complicated. I'm including it not because you absolutely need to know it in order to understand the everyday world around you but because the *f* orbitals look so freaking cool.

There are seven different types of *f* orbitals, some with six lobes and some with eight. You can see an image of the weirdest one in the above table, nicknamed "sausage with a double donut," which looks just like the p_z orbital wearing two inner tubes around its waist.

In an atom, seven *f* orbitals are stacked upon each other, which allows the molecule to minimize repulsions between the fourteen electrons (7 orbitals × 2 electrons per orbital = 14 electrons). But that requires forming an even crazier flowerlike structure. Because the *f* orbitals are mostly used in radioactive chemistry (and not in everyday chemistry), all you really need to know about them is that they have intricate shapes.

Regardless of the shape though, it's important to remember that each *s*, *p*, *d*, and *f* atomic orbital can only accommodate two electrons, and the electrons spin in opposite directions to minimize the interactions. And now that we know how electrons move within an atom, we can look deeper into how orbitals from *different* atoms overlap to form bonds and share electrons.

The first kind of bond that I'm going to discuss is called *head-on* overlap. This occurs when two orbitals overlap in one place.

Picture a typical Venn diagram that has three circles. If you remove one of the circles, you will be left with essentially two *s* orbitals. These two circles overlap in one place, which is exactly what happens when two *s* orbitals form a bond. The two spheres overlap with each other to form a single bond, which we refer to as a sigma bond.

When the sigma bond is formed, atom A's electrons now have a direct pathway to move toward atom B's protons (assuming that atom B is more electronegative than atom A).

But *s* orbitals aren't limited to bonding with their own kind. They can also form sigma bonds from head-on overlap with *p* orbitals. The new bond is formed when the *s* orbital overlaps with one lobe of the *p* orbital. If you take our two-circled Venn diagram and convert one of the circles to a figure eight, this image would model the bonding between an *s* orbital

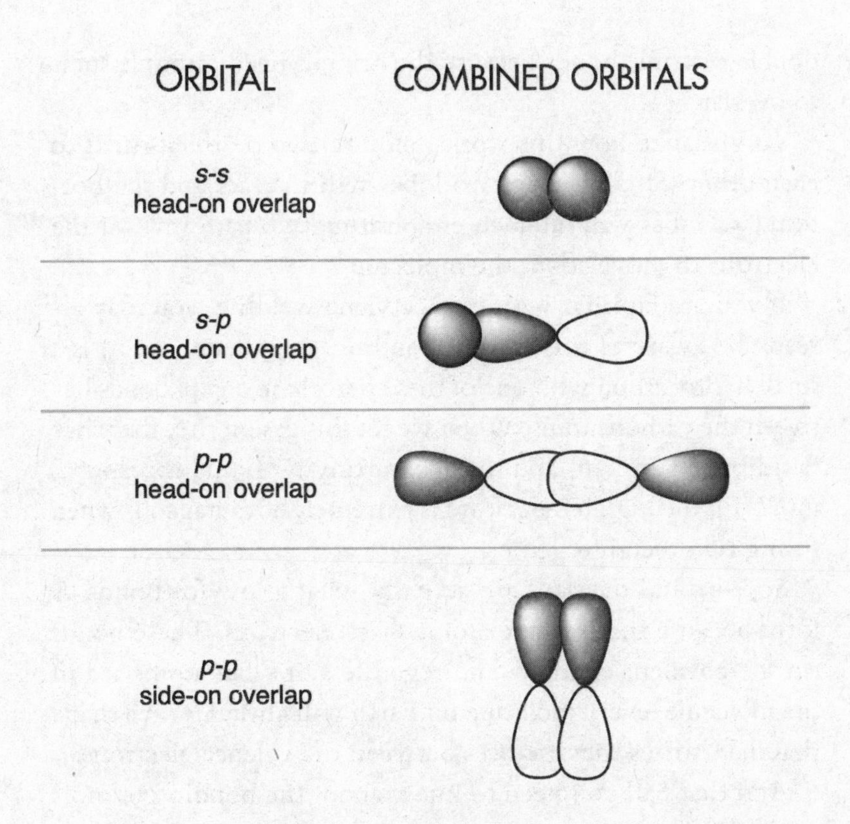

and a *p* orbital. There's one spot where the orbitals overlap, allowing for the electrons to easily move from one atom to the other.

Two *p* orbitals can also form a sigma bond if they interact through head-on orbital overlap. For this bond, the right lobe from the figure eight on the left must overlap with the left lobe from the figure eight on the right (∞∞). The orbitals overlap in one spot, producing a sigma bond.

However, two *p* orbitals can overlap with *side-on* overlap in addition to head-on overlap, which is exactly what it sounds like. A side-on overlap indicates that there are two places where the orbitals meet or are side by side (hence the name). This kind of bond is called a pi bond, and can only be formed through double or triple bonds because the orbitals need multiple spots to overlap.

To visualize how this works, picture two *p* orbitals next to each other (88). The top two lobes will interact and the bottom two lobes will interact, establishing two pathways for the electrons to move about the molecule.

If you are familiar with oxyacetylene welding, you may already be aware of this type of bonding. Acetylene (C_2H_2) is a small hydrocarbon with one of these fiercely strong pi bonds between the carbon atoms. When we set this gas on fire, the triple bond breaks in half, and the accompanying flame is 3,330°C (6026°F); the high temperature is extremely advantageous when fusing two metals together.

So—orbital overlaps are actually what allow for bonds to form because that's where atoms share electrons. These bonds can be covalent or ionic, and regardless of what atoms are in the molecule, every molecule on Earth will always form a shape that maximizes the distances between the valence electrons.

And that's all you need to know about the bonding in molecules. For starters, anyway!

Because now that you know how bonds are formed *within* molecules, I can share what happens when there are interactions *between* molecules.

Will two molecules react to form new ionic or covalent bonds?

Or will they ignore each other and just hang out in clusters?

3

LET'S GET PHYSICAL

Solids, Liquids, and Gases

In the last two chapters you learned about the foundation of chemistry: atoms and molecules. There are lots and lots of atoms in the world. Trillions of trillions! Lots and lots and lots and lots… You get my point. But we rarely see atoms and molecules out in the world, just hanging out. And that's partly because atoms and molecules are super-duper small, about a million times smaller than one human hair. Can you imagine how weird it would be to actually see the atoms that surround us? It would be extremely overwhelming.

Even if we *could* see atoms with the naked eye, we would be looking at groups instead of individual atoms. That's because atoms and molecules like to cluster together, like kids at an eighth-grade dance. When we're looking at charcoal for the grill, for example, we're looking at a cluster of carbon atoms. And when a bunch of carbon atoms and oxygen atoms decide to join together into a molecule of carbon dioxide, we can see that as solid dry ice.

In both of these examples, the atoms in the charcoal and the molecules in the dry ice are squeezed together, and there's barely any space between the atoms or molecules. This space is one of the main factors that determines what scientists call a *phase.*

In chemistry, there are three main phases: solid, liquid, and gas. (Other phases exist, like plasma and colloids, but I'm going to focus on what we see most often for now.) A super easy and often fun way to figure out whether something is a solid, liquid, or gas is to examine what happens when you drop the object.

A champagne glass, for example, will shatter with pieces of glass flying into a million different directions, landing at random. That's because the glass is in a solid state. It doesn't matter if the glass is broken—the pieces are still pieces. The glass didn't puddle together (like a liquid) or rise into the air (like a gas).

As it turns out, there are many intermediate phases of matter that don't fit neatly in the categories of solid, liquid, and gas. Glass is a solid. But more precisely, glass is an amorphous solid. This means that its physical properties are somewhere between a liquid and a solid. However, for the purpose of this discussion, we will assume that glass is just a typical solid.

When scientists look at a champagne glass under a microscope, we can see that the atoms are packed really closely together, like sardines in a can. The molecules are so squished together that they cannot move. They have a hard time rolling over or readjusting. Molecules in a solid state remind me of when my niece would fall asleep in my arms when she was a baby. It didn't matter what was happening around me, there was no way I was going to move and risk the chance of waking her. Nope, not happening. Molecules in a solid state are the same.

At the microscopic level, atoms in the solid phase are very similar to atoms in the liquid phase except for one major difference: the distance between the atoms. With a larger gap between the atoms, liquids can move about more freely, and take the shape of the container they're in. We see this whenever we accidentally drop that champagne glass. The glass shatters into pieces on a tile floor, but the liquid moves across the tiles until it reaches a corner or an edge.

In chemistry, we differentiate solids and liquids by discussing their shape and volume. Liquids are constantly changing their shape, but have a fixed volume. In contrast, solids have a fixed shape and volume. In the glass example, the champagne takes the shape of the glass, until the glass shatters and the champagne spreads across the floor. As a liquid, champagne has no definitive shape of its own.

Let's go through a couple examples of what I'm talking about. When you add a solid—say, a potato—to a container, it just sits at the bottom, right? It does nothing. And unless you're under the influence of something pretty strong, a potato in a pot definitely does not change its shape. However, when you add a liquid, like water, to that same container, the water spreads out, as much as it can, to cover the bottom of the container.

Remember that middle school dance? The molecules in liquids are slowly dancing on the dance floor, whereas the molecules in the solids are standing firmly in the corner. The liquids are sidestepping and waving their arms, while the solids have their feet glued to the ground. The liquids move to fill the container and the solids maintain their shape because their molecules are not dancing. In fact, the molecules in a solid are barely moving at all.

Most liquids on Earth are made up of molecules—except for two. At room temperature, bromine and mercury are the only liquids made from just *atoms*. All other liquids have at least one molecule. (For example, pure water is made of H_2O molecules—not just hydrogen or oxygen atoms, whereas pure liquid mercury is made of only Hg atoms.)

The difference between a liquid and a gas is the same as between a liquid and a solid—the distance between the atoms! Back to that middle school dance we go.

If the solids are standing still and the liquids are slow dancing, the gases are doing the quickstep. These molecules are moving as fast as they can, trying to spread out as much as possible. Unlike liquids and solids, gases do not have a definitive shape or volume. Instead, a gas tries to fill up the entire container. So, where a liquid would try to cover the bottom of a flask, a gas would attempt to fill it.

You're probably already familiar with common gases, like oxygen, nitrogen, and helium. Gases are moving around you right now (including inside your house) because our Earth's atmosphere is filled with gases. Even though we can't see oxygen or smell nitrogen or taste carbon dioxide, we wouldn't survive without these gases being present.

That's why we give astronauts spacesuits—because the moon and outer space do not have these gases the way the Earth's atmosphere does. It's also why scuba divers have to carry oxygen tanks on their backs. Without access to oxygen gas, humans die in approximately three minutes (as I'm sure you know).

But here on Earth, there are billions of molecules floating around you at this very moment. The majority of them are nitrogen (78%) and oxygen (21%). A whole whopping 1% of them are argon. And then there are trace amounts of a number of different gases (like carbon dioxide) and maybe even some pollutants (like carbon monoxide). When you take a deep breath in, you inhale the entire mixture of gases. The molecules travel through your nose and into your lungs before 4% of the oxygen is converted into carbon dioxide. When you exhale, you breathe out all of the nitrogen and argon molecules, about 17% oxygen, and 4% carbon dioxide. It is a common misconception that we exhale 100% carbon dioxide, and that's simply not true.

The argon that we exhale is an extremely stable gas. Scientists use argon whenever we need inert environments for our reactions. For example, when I was in graduate school, I used to pump argon gas into the flasks of my dangerous reactions to make sure that they did not catch on fire. The argon gas minimized the chance of an explosion, but I have to admit that conducting an experiment that could explode at any point is exhilaratingly terrifying.

Argon has the atomic number 18, which you now know means that argon has 18 protons in the nucleus and 18 electrons outside of the nucleus. Even though argon is relatively small, it is very dense—or packed tightly together in a small space.

When I teach students about gases at UT, I like to use balloons filled with argon and helium to show how the density of gases matter. I hold the argon balloon, which looks like a regular balloon, and then I drop it. The balloon immediately sinks to the ground because argon is heavier than air. Then I release the balloon filled with helium, and it immediately floats to the ceiling. That's gas density in a nutshell.

Dense gases have more molecules packed together in a set volume. If laundry were molecules in a gas, I would expect for a college student's laundry hamper to be "dense" because it is likely to be filled to the brim with dirty clothing. Marie Kondo, on the other hand, would have a significantly less dense laundry bin because she only has clothes that spark joy (and probably keeps up on her laundry better than a college student).

Gases that are less dense, like hydrogen and helium, float away because the gases are lighter than air. These lighter gases are perfect for the celebratory balloons we discussed in the last chapter, but—as you've probably learned—you need to tie or weigh the balloon down to keep it from flying away.

But how does something like helium transition from a gas to a liquid or from a liquid to a solid? These types of phase changes, which you may remember from high school, happen around you on a daily basis. Melting, vaporization, condensation, and freezing are all processes that are a direct result of increasing or decreasing the distances between the molecules in a given substance.

The easiest phase change to start with is melting. I don't know about you, but I learned about melting at a very young age. I was outside, eating ice cream, and the hot sun beating down caused the ice cream to drip down my hand. It was a total mess, and a horrible introduction to one of the main phase changes in chemistry. The funny thing is, "melting" is actually the wrong word for this type of science. The technical term is "fusion," but nobody ever says that.

When ice cream—or anything else—is fusing or melting, the distances between molecules become larger and larger, causing the solid to turn into a liquid. So, if the molecules in a solid are—I'm going to use an exaggerated number here—one mile apart, the molecules in a liquid are now five miles apart. Realistically, atoms in solids are about 10^{-10} meters apart, but I find that number to be very hard to visualize.

Here's the important part to remember: after a phase transition, the molecules are still the exact same molecules. The atoms and the distances between the atoms have not changed, but the molecules themselves are farther apart.

But how do the distances increase? They need a source of energy, usually in the form of heat. If we change the temperature of the environment, we can force the molecules to speed up (from heat) or slow down (from cold). As you will see in just a bit, this also affects the distances between the molecules.

This should seem fairly logical if you think about our ice cream example. In order for the ice cream to melt, it needs an external heat source. In Texas, if I eat ice cream outside, it will start to melt after just a few minutes. The heat from the molecules in the atmosphere provides enough energy that the molecules in the ice cream begin to move, ultimately increasing the distances between the molecules. When this process happens microscopically, the ice cream begins to melt, causing fusion to occur.

The ultimate example of fusion is the first step in the process of making chocolate-covered pretzels: melting the chocolate. When I do this at home, I like to place the chocolate in a heat-safe bowl and rest it over a pan of boiling water. This setup allows the heat from the steam to transfer through the bottom of the bowl, pumping directly into the chocolate. The extra energy forces the molecules in the chocolate to start to wiggle around, essentially allowing the molecules to increase the distances between each other. I know the exact moment this happens because I can watch the chocolate begin to melt before my eyes.

When I remove the melted chocolate from the pan, I can see another physical change. The water in the pan is boiling because enough heat has been added to encourage the liquid water to convert into steam. And as the water in the pan becomes steam, the space between the water molecules greatly increases. Therefore, if the molecules in a solid and a liquid are one mile and five miles apart, respectively, then the molecules in a gas would be about fifty miles apart. Again, the molecules have not changed, but are simply much, much farther apart in a gas than they would be in a liquid or solid. We already know that gases do not have a definitive shape or volume, so the gaseous water molecules—or steam—rise into the air and seemingly disappear.

This process of converting a liquid into a gas is called vaporization, though many people incorrectly call it evaporation. This is a common misunderstanding, but let's discuss the differences. Like fusion, the process of vaporization increases the distances between molecules, which means that heat is required for it to occur. This happens at the boiling point of the liquid, which is the temperature at which the liquid turns into a gas.

Evaporation, on the other hand, has to do with how the molecules convert from liquids to gases without directly adding a lot of heat. This phase change happens below the boiling point, like how a glass of water evaporates overnight or how sweat evaporates off your body. Neither of these processes require a blowtorch—the molecules simply have enough energy to transition to a gas. Conversely, the boiling water was given more energy to help it change into a gas.

In either case, we can only convert a liquid into a gas by increasing the distances between the molecules. Readers who are bakers experience what happens when boiling water turns into gas while they're melting chocolate. But have you ever had your melted chocolate seize on you? The pesky gaseous water molecules can get stuck in melted chocolate and cause another phase change: condensation.

When the water molecules *condense*, it causes the smooth chocolate to seize into a grainy, disgusting mess. During this process, the gaseous water molecules (steam) turn into liquid water molecules and interfere with what's happening at a molecular level with the chocolate. This phase change is the same one that happens when water droplets form on the outside of your beverage on a hot day.

Condensation and vaporization are equal but opposite processes. It's like my commute to work—the distance and time spent is the same. I drive ten minutes on my way there and ten minutes on my way back home. But what differs is the direction I'm going. Similarly, while vaporization *increases* the distance between molecules, condensation *decreases* the distance between the molecules. And that in turn allows a gas to transform into a liquid by forming attractions between neighboring molecules.

Liquids can also change into solids without changing chemical composition. This process is called freezing, and it occurs when molecules get close enough together to convert from a liquid to a solid. And just how vaporization and condensation are opposite processes, freezing and fusion (what we think of as melting) are opposite processes. Fusion requires the molecules to spread out and increase the distances between each other, in order to convert a solid to a liquid. But for freezing to happen, the molecules need to be close together, allowing the substance to transition from a liquid to a solid.

The best way to freeze something is to stick it into a cold environment, like a freezer, but you could also change its pressure (in a lab). The low temperatures force the molecules to slow down, ultimately decreasing the distances between the molecules. When I put my chocolate-covered pretzels into the freezer, the melted chocolate solidifies into a hardened chocolate casing. The process doesn't happen immediately and is dependent on the thickness of the chocolate coating. The more molecules there are, the longer it will take for them to slow down enough to form the solid. But in general, all molecules have a freezing point, which is the temperature at which the liquid turns into the solid.

Fusion, vaporization, condensation, and freezing are the most common phase changes, but there are two more physical transitions that are less common but worth mentioning: sublimation and deposition. These changes occur when solids turn directly into gases and when gases turn directly into solids. Molecules never turn into a liquid during sublimation or deposition, they just go from solid to gas and gas to solid. For these transitions to happen, the distances between the molecules need to increase or decrease quickly and drastically. Depending on the molecule, these two phase changes can happen naturally in a classroom or in a lab under extreme temperatures and pressures.

Sublimation doesn't happen all that frequently in nature because molecules have to move too quickly. In fact, in our daily lives we don't experience it all that much. Most of us only interact with these phases when we handle dry ice. Dry ice (or solid carbon dioxide) has unique properties that allow it to transition from solid to gas, meaning that the distances between the molecules rapidly increase during the phase change. This process happens spontaneously at atmospheric pressure and temperature, which is why dry ice is commonly used to produce fog in musicals or at concerts—and in my classroom.

Sublimation is also used in air fresheners and mothballs. As solids, these substances release a small puff of molecules into the atmosphere over time and create odors. Each system sublimes at room temperature, but unlike dry ice, the process can take days if not weeks to complete. That's why your car air freshener needs to be replaced every few weeks—the molecules in it stop subliming into the air.

The opposite of sublimation is deposition, where the molecules in the gas phase convert directly into a solid. Here, so much energy is lost during the transition that the molecules essentially stop moving wherever they are and just sit down. For those of you that live in colder climates, you probably experience deposition more than you know. Every morning, when you look outside to see frost-covered leaves, you are looking at the result of deposition. The water molecules in the air lose so much energy at night that they just deposit themselves on the leaves to form a beautiful icy wonderland. If you ever decide to sit outside and watch frost form, you will see it go straight from water vapor to solid ice without transitioning through liquid water.

Another common example of deposition is the soot that forms on the inside of a chimney. When I lived in Michigan, I loved to sit by the fireplace on cold mornings and soak up the heat, maybe even with a hot cocoa. I didn't know it at the time, but had I been paying attention, I would have been able to observe the soot particles combining with dust as they converted from the gas phase to the solid phase. These soot/dust particles would collect on the inside of the fireplace, and leave behind a black filth that my mother hated. In this context, the soot deposition happens on a much faster timescale than frost deposition, but in my biased opinion, they are equally mesmerizing.

To recap, there are six kinds of phase changes. I've summarized all of them in this table.

NAME	PHASE CHANGE
Fusion/Melting	solid → liquid
Freezing	liquid → solid
Vaporization	liquid → gas
Condensation	gas → liquid
Sublimation	solid → gas
Deposition	gas → solid

Most molecules have specific temperatures and pressures that will allow them to go through all six phase changes, but every molecule is unique. Some of them even have a *triple point*, which is a specific temperature and pressure where the distances between the molecules are so ambiguous and undefined that the substance will exist in the solid, liquid, and gas phase—all at once. For water, the triple point occurs at a temperature of 0.01°C (32°F) and a pressure of 4.58 torr. The most common way to observe it in a lab is by trapping water in a closed container and putting it under a vacuum to decrease the pressure.

However, have you seen those YouTube videos where someone throws a pitcher of boiling water into the air at –52°C in Alaska? As soon as the water leaves the pitcher, it changes phase: some of the water molecules instantly freeze into small icicles, but the remaining molecules vaporize into a huge cloud of white gas. It looks like a frozen firework, where there is a big puff of white gas with cool icicles arching toward the ground. All three phases of water exist (for just a second) at one time. That's kind of what water is like at its triple point, and it's freaking awesome.

There's another set of conditions, again at a specific temperature and pressure, that indicate the last moment where the liquids and gases can be distinguished. When you go above what's called the *critical point*, the distances between the molecules in the liquids and gases are too variable for us to define the material as either a liquid or a gas. Instead, we refer to it as a *supercritical fluid*, which is a weird liquid-gas type of substance. It has some properties of a liquid and some properties of a gas (the properties vary for different types of molecules).

One of the most common uses of a supercritical fluid is during the decaffeination of coffee. The coffee beans are steamed before being pumped into a special container that can withstand high pressure conditions. At this point, supercritical carbon dioxide is sprayed over the coffee beans, dissolving the caffeine in the liquid-gas material. The beans themselves are not vulnerable to the supercritical fluid, which makes it the perfect solvent to extract the caffeine. The coolest part of this process is that the caffeine can be removed from the supercritical carbon dioxide, allowing for the solvent to be used over and over again.

Supercritical carbon dioxide also used to be a preferred solvent for some dry cleaners because it could easily remove dirt from the clothing without actually getting the clothing "wet." (I'm using quotation marks here because a supercritical fluid does not follow the typical definition for wetness. The liquid/gas material is not exactly wet, but it's definitely not dry.) The material was sprayed on the clothing under pressurized conditions, but there was one major problem. When the pressure was released, some fragile buttons would shatter or pop off the clothing. They were unable to perfect the super-freezing process of clothing so most cleaners currently forgo this treatment for alternative options.

But all of these phase changes that I just mentioned are on the macroscopic level. We can see condensation, freezing, and even supercritical fluids with the naked eye. But what we cannot see is *how* all of these changes occur—because that happens at the microscopic level.

HOW SCIENTISTS "SEE" THE WORLD

When chemists or biologists or geologists or any science-ologist studies the world, we're considering two different perspectives: the macroscopic (what we can see) and the microscopic (what we can't).

If you have to get a microscope out to see something, it's *microscopic.*

If you can see it with the naked eye, it's *macroscopic.*

So, what is going on among the teeny tiny molecules? The first thing chemists look for is how the electrons are distributed within the molecules, which is determined by—you guessed it—the shape of the molecule. That's because the shape of a molecule tells chemists like me *how* the electrons from different molecules interact with one another, and more importantly, how they will arrange themselves in space.

In some systems, the molecules are lined up neatly like members of a conga line, while others take more of the head-to-toe configuration like a yin-yang symbol. It's relatively easy to identify the most common patterns of molecular arrangements, and this will finally explain *how* groups of molecules change from phase to phase on the microscopic level.

But in order to do that, we first need to determine the overall polarity of the molecule.

And that comes back to a topic you are already familiar with. Electronegativity.

Consider oxygen, which is one of the most electronegative atoms. Remember that means when oxygen is in a molecule, it is going to suck all of the electrons from neighboring atoms toward its nucleus. The oxygen atom in the water molecule (H_2O) will cause all of the electrons to hover near the oxygen—not either of the hydrogens.

Since the electrons are unevenly sitting on the oxygen side of the molecule, we give the oxygen a partial negative charge. This is exactly what we did when we were looking at how atoms share electrons within a bond. But now, we are going to look at what happens when you put multiple bonds together in one molecule.

There are two different ways that the electrons can distribute themselves across a molecule, giving us polar and nonpolar molecules. If the molecule can be split in half symmetrically, then it is considered to be a *polar* molecule. This means that electrons are not perfectly distributed across a molecule. Instead, there is a positive side and a negative side—just like a standard magnet.

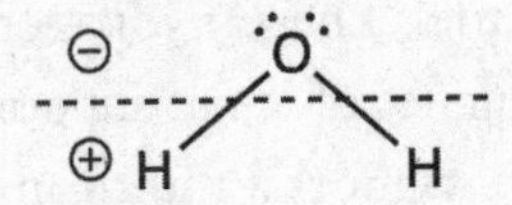

Let's look more closely at how the electrons are distributed in water. Like I mentioned earlier, the oxygen in water carries a partially negative charge. Therefore, both hydrogens carry partially positive charges. This is true for every single water molecule on Earth. The oxygen is always partially negative and the hydrogens are always partially positive. In these situations, we can actually divide the molecule in half to give us one positive side and one negative side, creating poles on the molecule.

These polar molecules cause a chain reaction of strong attractions between the positive side of one water molecule and the negative side of another water molecule. These strong attractions between molecules are called dipole-dipole interactions. Dipole-dipole interactions are limited to molecules with a permanent imbalance of charge (i.e., *polar* molecules).

Hundreds of dipole-dipole interactions are happening around you at this very moment. If you are in your kitchen, they are in your apples and pears; they are even in your pork, beef, and fish. And if you have a glass of water, soda, or even some wine nearby, you are also interacting with an extra special set of dipole-dipole interactions that are so strong that we give them their own name. These attractions are called hydrogen bonds, and they are ridiculously robust. Water molecules are perfect examples of molecules that exhibit hydrogen bonding. Why? Because they are polar molecules that contain extremely polar bonds.

But keep in mind that hydrogen bonding is not the covalent bond that's happening when hydrogen and oxygen atoms come together to make H_2O. Hydrogen bonding occurs between a hydrogen atom on one water molecule and an oxygen atom on another water molecule. These hydrogen bonds are *so* strong that it only takes six inches of ice to support a loaded semitruck.

Six inches of ice to support a multiton truck! Crazy, right?

There used to be a show called *Ice Road Truckers* that I was obsessed with, and it was the perfect example of hydrogen bonding. As a native Michigander, I know all too well that thin ice is extremely dangerous, and I could barely stomach watching these brave truckers drive across miles of frozen ice. But hydrogen bonds are so freaky strong that a heavily loaded truck could drive across the frozen lakes in Canada.

Luckily, the truckers have intricate ways of assessing the integrity of the ice, in order to avoid suffering a horrible accident. But what they might not know is that they are actually investigating the strength of the attractions that exist between the water molecules. See, when these hydrogen bonds break, the molecules are able to go through a phase change.

When a small portion of the hydrogen bonds are broken, the frozen ice can melt into liquid water, creating a serious problem for anyone messing around on frozen lakes. But when *all* of the hydrogen bonds break, the liquid water can then transition into water vapor (or steam). So, when we watch an ice cube melt or a pot of water boil, we are actually observing hydrogen bonds break apart in real time.

Conversely, we can watch hydrogen bonds form when an ice cube freezes and liquid water turns into solid water (ice). I take advantage of this phase change every time I perform one of my biggest demonstrations called the thundercloud. When I throw hot water into a bucket of liquid nitrogen, I force the water to freeze at the bottom of the bucket. During this process, the heat from the hot water is transferred to the liquid nitrogen, causing the liquid nitrogen (N_2) to vaporize into a huge cloud of nitrogen gas.

Like water, the attractions between the nitrogen molecules

have to break before the nitrogen can change from liquid to gas. But unlike water, nitrogen cannot form hydrogen bonds because these attractions are limited to extremely polar molecules. Instead, nitrogen molecules form dispersion forces between themselves.

Dispersion forces occur when there's a relatively weak interaction between the molecules in a given sample. Remember those pesky trans fats that we discussed in the previous chapter? The reason they are able to stack together so well (and clog our arteries) is because they use dispersion forces to lock the molecules tightly together. This is true for any nonpolar molecule.

So, what exactly does it mean to be a *nonpolar* molecule?

Nonpolar molecules do not have a positive side or a negative side. Instead, all of the electrons are symmetrically spread throughout the molecule—just like how a perfect chocolate chip cookie has an even distribution of chocolate chips across the entire cookie. It would not be possible to split the cookie in half to have one side with more chocolate chips than the other. The same thing goes for a nonpolar species—the electrons are evenly distributed across the molecule.

But here's what is really neat about nonpolar molecules: for a nanosecond, they can become polar molecules, but then they immediately go back to normal. Just like how I can play dress up in a photo booth for a few minutes, but then I can take off the silly hat and glasses to return to regular old Kate.

So how do molecules play "dress up" to get an uneven distribution of electrons? Well, every atom and molecule, regardless of size, can have a moment when the electrons are unbalanced in the atom. For example, the nitrogen molecule (N_2) has fourteen electrons shared between the two nitrogen atoms. It's possible that, for a split second, six electrons will be on the left side of the molecule and eight electrons are on the right. In that moment, the nitrogen molecule has a very small partially positive charge on the left side of the molecule and partially negative charge on the right side of the molecule.

In my thundercloud experiment, one nitrogen molecule (molecule A) is sitting very close to another nitrogen molecule (molecule B). When eight electrons suddenly appear on the right side of molecule A, the electrons in molecule B will feel that negative charge and run away from it. This is kind of like being at a haunted house with your friends when a skeleton jumps out from nowhere. You (and your friends) all jump back and sprint in the other direction, right? Well, that happens with dispersion forces too. Just one moment of unbalanced charge in a single molecule—or just one skeleton to scare a large group of friends—triggers a domino effect of charge through an entire species (or bunch) of molecules.

That being said, a molecule will attempt to redistribute its electrons as soon as it can, in order to achieve its constant mission of finding the greatest distance between its electrons. The effect lasts for less than a second before the domino effect happens all over again. This cascading effect of electrons is very common, and it is the reason why nonpolar molecules can remain together in clumps instead of floating off into the atmosphere. Without these interactions, each liquid nitrogen molecule would separate from its neighboring molecule, vaporize, and then drift away into outer space, ultimately ruining my big experiment.

These attractions between molecules are so common (and important) that they have been given their own name: intermolecular forces (IMFs). Hydrogen bonds, dipole-dipole, and dispersion forces are all types of IMFs. When these attractions are formed between molecules, gases can turn into liquids and then liquids can turn into solids. On the other hand, when these attractions are broken, solids can turn into liquids and then liquids can turn into gases.

For the thundercloud experiment, I form hydrogen bonds between the water molecules when the water freezes and then I break dispersion forces between the nitrogen molecules when the nitrogen vaporizes. These two physical changes happen so quickly (and in a confined space) that I can produce a show-stopping three-story-tall cloud.

As you may have noticed, I find phase changes and IMFs to be absolutely fascinating. I could write for days about how the distances between molecules and their corresponding IMFs dictate what phase a substance ends up in. But I'm guessing you're feeling ready to move on—so how about we blow some stuff up?

4

BONDS ARE MEANT TO BE BROKEN

Chemical Reactions

So far, we've covered atoms and molecules and phase changes. We've learned how water is made of two hydrogen atoms and one oxygen atom, and can be a solid (ice), a liquid (what comes out of the tap), and a gas (steam). But what happens when another molecule comes in—a completely different molecule—and breaks the bond between hydrogen and oxygen that creates H_2O? Are the atoms rearranged to form new molecules? If new molecules are produced, can we reverse the reaction and bring the original molecules back? Or is it like Marty McFly in *Back to the Future*, where one little difference changes everything?

These questions are my favorite part of chemistry because the answers are the basis for chemical reactions.

There are two concepts you need to know before we can dive into reactions. First is the difference between a chemical equation and a chemical reaction. Confusing the two is like nails on a chalkboard to a scientist. Or your college professor (ahem).

Luckily, the difference is pretty darn simple.

A chemical *reaction* happens in a lab.

A chemical *equation* is written down on paper.

In the lab, I can perform a chemical reaction by mixing two chemicals together in a flask. I'm usually wearing a lab coat and carefully observing each step of the chemical reaction. During this process, the reaction may change color or even phases (like from a solid to a liquid) because stuff is happening at the molecular level.

Atoms are being rearranged.

In contrast, if I wanted simply to document the experiment by highlighting which chemicals were used and how much of each, I would do this by writing out what's called a chemical equation, which has three distinct pieces: (1) a reactant side, (2) an arrow, and (3) a product side. The reactants are always on the left side of the arrow and the products are always on the right side of the arrow. A generic chemical equation looks like this:

$$\text{Reactants} \rightarrow \text{Products}$$

Or like this:

$$A + B + C \rightarrow D$$

A, B, C and D represent different molecules, like water or carbon dioxide. But let's think about it in terms of something more interesting, like dessert. If my chemical reaction is the process of making a cake, my reactants would be all of the chemicals—or ingredients—that I need to make the cake. So, in my chemical equation, all of my ingredients (e.g., flour, sugar, eggs) would be on the left side of the equation. The products would then be all of the chemicals that are produced during the chemical reaction—the cake! Therefore, the chemical equation for making a cake would look something like this:

Flour + Eggs + Sugar → Cake

The chemical equation as written above indicates that the cake recipe is one part flour, one part eggs, and one part sugar—or one cup of flour, one egg, and one cup of sugar. If you bake, you're probably screaming that this is a horrible cake recipe. Because these are terrible ratios for a traditional cake, this chemical reaction would create an end product that tastes awful.

When the ratios of chemicals are wrong in a chemical equation, we say that the equation is unbalanced. Calling an equation unbalanced indicates that we have a bad recipe, which will give us a bad product. In chemistry, unbalanced chemical equations are useless. In these situations, we have to balance our chemical equation. We can do this by adding coefficients—or numbers—to the chemical equation. These coefficients are added before the molecule in the chemical equation to indicate the proper ratio needed to make the product. If we need three cups of flour, four eggs, and one cup of sugar to make a cake, we adjust our chemical equation to indicate these quantitates:

3 Flour + 4 Eggs + Sugar → Cake

Note that the number 1 is not used in the chemical equation. All 1 coefficients are implied and not included in any chemical equations.

We could easily adjust the recipe to make chocolate cake just by adding one more reactant, cocoa powder, which gives us:

3 Flour + 4 Eggs + Sugar + Cocoa Powder → Chocolate Cake

This equation is also unbalanced because cocoa powder is really bitter, which means we have to adjust the amount of sugar needed. Therefore, we balance the equation by adjusting it. The new recipe could be:

$$3\ \text{Flour} + 4\ \text{Eggs} + 2\ \text{Sugar} + \text{Cocoa Powder} \rightarrow \text{Chocolate Cake}$$

Even a recipe for chocolate cake could quickly be manipulated to make brownies or chocolate cookies. This is because flour, eggs, and sugar are the building blocks for many desserts, just like atoms and molecules are the building blocks for chemistry.

Let's go back to our generic equation:

$$3A + 4B + C \rightarrow D$$

This chemical equation is loaded with valuable information. It provides me with a protocol—or a recipe—to follow so that I can make exactly one product, D. And if I wanted to make exactly one D, I would add three parts of A, four parts of B, and one part of C to a flask. I would stir them for a few hours, maybe add some heat, and eventually one part of D would form.

But what does "one part" of D mean? Is it one cup? One gram? One kilogram?

Actually, it's one *mole.*

You might be thinking, wtf is a mole? In chemistry, a mole is not a cute furry animal or a delicious chocolate sauce. Instead, it is a very specific number that helps us to identify how many molecules are in a reaction. This brings me to the second thing you need to know in order to understand chemical reactions—namely, what a mole is and why it's important.

The idea of the mole was first proposed back in 1811 by an Italian scientist named Amedeo Avogadro. However, it was

German chemist Wilhelm Ostwald that first used the term "mole"—short for the German word *Molekül* (molecule).

Without using the word *mole*, Avogadro suggested that if two samples of gas have the exact same temperature, pressure, and volume, then they will have the exact same number of molecules. The identity of the gas was irrelevant IF the three conditions matched.

For example, let's say I have a balloon of oxygen gas and a balloon of nitrogen gas in my classroom. They are both at the same temperature, and they are the exact same size. Since the volume of the balloon is not changing, this means that the pressure inside the balloon is the same as the pressure outside of the balloon. It also means that the pressures of the two balloons are the same. If the temperature, volume, and pressure of the balloons are identical, then Avogadro proposed that the balloons will have the exact same number of molecules inside the balloons. In other words, my nitrogen balloon has the exact same number of molecules as my oxygen balloon. The only difference is the type of molecule that is in each balloon.

Austrian chemist Josef Loschmidt was able to quantify the number of molecules in a gaseous sample in 1865 when he proposed an equation for number density—or a way to calculate the number of molecules in a given volume. He discovered a very unique constant that supported everything Avogadro originally proposed back in the early 1800s. Therefore, in 1909, when French physicist Jean Perrin used Loschmidt's "magical" number to convert the mass of a sample into the corresponding number of molecules, he referred to it as Avogadro's Number as a way to honor Amedeo for his original work on the topic.

I always wondered if Loschmidt felt slighted by the naming of the infamous number. Regardless, Perrin defined

6.022×10^{23} as Avogadro's Number, which represented the exact number of molecules in a 32 g sample of diatomic oxygen.

Perrin's discovery was groundbreaking at the time. But in 2019, the mole was redefined. The International Union of Pure and Applied Chemistry (IUPAC) wanted to adopt simpler definitions of a few of the base units; therefore they proposed a new definition of the mole. It was readily accepted because we no longer have to compare the number of atoms to a specific sample, like carbon or oxygen.

In the new definition, a mole is defined as a sample that contains exactly 6.022×10^{23} entities. As a chemistry professor, I did a little dance when I heard this new definition. It is *much* easier to teach that the mole is just a number instead of forcing my students to memorize the entire history of Avogadro, Loschmidt, and Perrin.

With the new definition, the word *mole* indicates the number 6.022×10^{23}. That's it. It's just a number. Just like how a decade means 10, a century means 100, and a gross means 144. A mole means 6.022×10^{23}.

Remember that whole thing in the last chapter about the world we can see (macroscopic) and the world we can't (microscopic)? Moles bridge that gap. We use moles to convert mass from the macro world to the number of molecules in the micro world.

Moles rule when scientists like me need to determine the number of molecules in a given sample, which we do, when we're making a cake or creating an explosion.

A mole in chemistry is gigantic. To give you a reference point, 10^6 is a million, 10^9 is a billion, and 10^{12} is a trillion. The actual value for a mole is 602 sextillion or 602,200,000,000,000,000,000,000.

602,200,000,000,000,000,000,000!

MOLES ARE NOT GRAMS (OR TEASPOONS OR TABLESPOONS OR PI)

Keep in mind that 3 moles of A, 4 moles of B, and 1 mole of C is not equivalent to 3 grams of A, 4 grams of B, and 1 gram of C. Moles don't work that way. Remember atomic mass from the periodic table? The atomic mass not only indicates the average number of protons and neutrons, but it also tells us how many grams of each element exist in one mole.

Take cobalt for instance. If we look at the periodic table at the back of this book, we can see that there are 58.93 grams of cobalt in one mole of cobalt. So, if my chemical equation calls for 3 moles of cobalt, I know that I need to measure out 176.79 grams of cobalt (58.93 × 3 = 176.79) in my lab. If I only added 3 grams of cobalt to my reaction, it would not work very well because I am missing 173.79 grams of starting material.

We use moles in chemical equations to make sure that we have the perfect ratio of atoms needed for the chemical reaction. Otherwise, it would be like mixing six buckets of flour and one cup of sugar, in an attempt to make a birthday cake. It just won't work.

Daniel Dulek, a pediatric infectious disease specialist, did a TED-Ed talk on the mole and provided the best analogy of a mole that I've ever heard. If you were given a mole of pennies on the day you were born, and then threw away a million dollars *every second* until you turned a hundred years old, you would still have 99.99% of your money on your hundredth birthday.

In one hundred years, spending a million dollars *every second*, you would have only spent 0.01% of your money.

Can you believe that?

A mole is a freaking huge number.

But to come back to the original point, we use the unit mole to provide the ratios of molecules needed for chemical reactions. The number of moles is indicated by the coefficient in the chemical equation.

So if we say that we need 3 moles of A, 4 moles of B, and 1 mole of C to produce 1 mole of D, this really means that we need 1.807×10^{24} molecules of A, 2.409×10^{24} molecules of B, and 6.022×10^{23} molecules of C to form 6.022×10^{23} molecules of D. (Remember, 1 mole is equal to 6.022×10^{23} molecules, therefore 3 moles of A is really 1.807×10^{24} molecules or $6.022 \times 10^{23} \times 3$.)

However, it is much easier to represent all of that information in the below chemical equation:

$$3A + 4B + C \rightarrow D$$

Now that you understand the moles and equations specific to chemistry, we can get to the good stuff: examining the different types of chemical reactions.

When we look at typical chemical reactions, we are breaking and making bonds, a process that is directly correlated to the addition and release of energy. This branch of chemistry is called thermodynamics—you might have heard it before in relation to warming or cooling technology. But what you need to know for the purposes of this chapter is that thermodynamics is all about the energy flow of a chemical reaction.

Energy flow is either positive or negative. We calculate the energy flow by looking at the total energy required to break all the bonds and the total energy released when all the bonds are formed. The easiest way to remember it is:

$$\text{Total Energy} = \text{Bonds Broken} - \text{Bonds Formed}$$

If more energy is put into the reaction than released, then the total energy for the reaction is positive. Let's play with some bullshit numbers to help us take a deeper look at this concept. (For this example, I will use joules, which is one of the most common units of energy. In chemistry, the range of energy values is such that we typically use kilojoules (kJ) as a unit of measurement. The prefix kilo indicates we're referring to a thousand joules.)

In this instance, let's say we need 500 kJ of energy to break all the previous bonds and release 250 kJ of energy when forming the new molecules. The equation would look like this:

total energy = 500 kJ – 250 kJ

total energy = +250 kJ

The net energy charge is positive, or +250 kJ. In this example, more energy was needed to break the bonds than released when forming the new bonds. This is likely because the bonds in the original molecule were stronger than whatever just formed. Because the reactants (the previous bonds) were more stable or positive than the products (the new bonds), the energy changes are defined as *endothermic.*

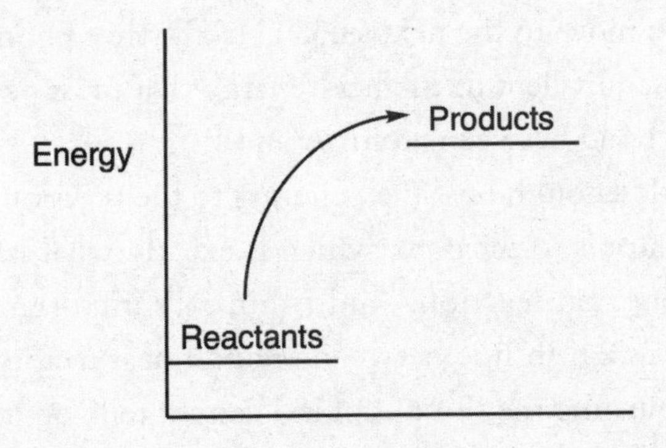

Whenever we break a bond in a chemical reaction, we need to add energy. This means the process of breaking bonds will always be endothermic. Let's look at how this works in the below equation where a generic covalent bond, A–B, is broken into atom A and atom B:

$$A\text{–}B + \text{energy} \rightarrow A + B$$

We add the term + *energy* to show that this is an endothermic process.

These chemical reactions operate exactly like the game Red Rover that we used to play as kids. Do you remember that game? One team would hold hands and one person on the other team would have to run between two people to try to break their connection—or their bond. The two people never just drop each other's hands during the game. Someone has to run with enough energy to break the connection between person A and person B. This is similar to how we can break the bond between atoms A and B. We have to do something to encourage the atoms to separate. The bond just won't break on its own.

To understand this entire process, just think about what happens when you walk upstairs. When you move from the bottom to the top of the stairs, you have to use your energy to pick up your leg to move to the next step. The effort we put in to walk upstairs is equivalent to the heat/energy that must be added to break the bond between atoms A and B.

If we add enough heat (i.e., energy) to the reaction, you can force the atoms to separate, which is exactly what happens in a decomposition reaction. And I think it's important to note that there is a thin line between enough heat to force a reaction and burning the shit out of it. I cannot tell you how many times I've burned samples in the lab and cookies at home. And just like when you burn food, the decomposition reaction will cause the molecule to turn black. Maybe even release a bad odor.

Molecules like aluminum hydroxide decompose really quickly when enough heat has been added to the system. Their bonds break instantly, and the atoms separate as a result. During decomposition, it absorbs so much heat that it protects whatever is underneath it. That's why it is usually used as a filler in flame-retardant materials (because the heat cannot penetrate the layer of aluminum hydroxide). As you might expect, I'm a big fan of this compound because of its strong endothermic properties.

Other molecules require even more energy to break the bonds within the compound. For example, when molecules, like oxygen, interact with high energy like UV radiation, the bonds in the molecules dissociate, or break apart. The UV energy is so strong that the molecule immediately breaks into different pieces. When this happens to the oxygen gas that we breathe—what we refer to as diatomic oxygen, O_2—the double bond breaks, releasing two monoatomic oxygen atoms (O) as shown below:

$$O{=}O \rightarrow O + O$$

This decomposition of oxygen can only occur when the incoming energy is absorbed by the molecule. It breaks the double bond and forces the two oxygen atoms to move to a higher energy state. When this reaction happens in the stratosphere, the two oxygen atoms are so dissatisfied with the situation that they instantly try to reform the double bond. Some monoatomic atoms even grab a third oxygen to produce ozone (O_3). The atoms will basically do anything possible to reform a bond with neighboring atoms.

So, how does that process work? What is the chemistry behind making bonds?

To answer these questions, let's go back to our bullshit numbers again. We know that we need +500 kJ of energy to break the existing bonds, but let's pretend that the reaction releases 750 kJ of energy when forming the new molecules. The net energy change was –250 kJ this time, which means that more energy was released than absorbed during this chemical reaction.

Total Energy = Bonds Broken – Bonds Formed

total energy = 500 kJ – 750 kJ

total energy = –250 kJ

When the new bonds are stronger than the original bonds, the reaction is *exothermic.* And the cool thing about negative energy reactions is that they often happen on their own.

If we look at an exothermic chemical reaction between solid barium metal and chlorine gas, we will see how two species can come together to form a new bond. In this case, the barium metal forms an ionic bond with the chlorine gas, resulting in the new ionic molecule, barium chloride. This chemical reaction can be represented by the chemical equation shown below:

$$Ba + Cl_2 \rightarrow BaCl_2$$

Even though this equation might not seem to show it directly, trust me when I say that both barium and chlorine will end up in a place of lower energy when they form the ionic bond with each other. When the bond is formed, energy is released because the reactants start at higher energy than the products.

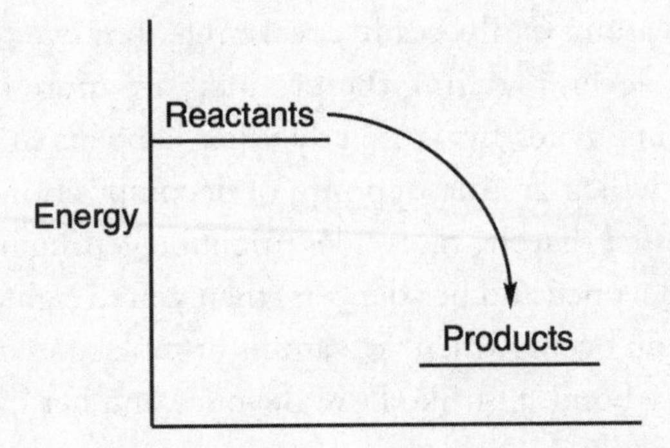

We can represent that in the chemical equation like this:

$$Ba + Cl_2 \rightarrow BaCl_2 + \text{energy}$$

Or more generically as:

$$A + B \rightarrow A\text{–}B + \text{energy}$$

When a bond is formed between two atoms, the energy within those atoms lowers. Nature is always trying to get to a place of lower energy; just like it's exhausting for us to do pull-ups after a long day, atoms also don't like to go to higher energy in a chemical reaction. Lower energy is a good thing in chemistry because the resulting molecule is much more stable than the original atoms.

Remember, in chemistry stability means that the molecule is less likely to react with other molecules—more importantly, that the molecule's electrons are being pulled toward the protons in each atom's nucleus. This strong attraction between the electrons and protons means that the valence electrons are more "protected," therefore they have a harder time reacting with *other* molecules.

This is what happens when the overall energy change is negative. The atoms easily, readily rearrange to move to the lowest energy level, meaning the products are more stable than the reactants. And this is exactly what happens in formation reactions, which are the opposite of decomposition reactions. If you guessed that the molecules might be getting into formation like Beyoncé and her dancers, then you're right. A formation reaction occurs when two atoms or molecules combine to form a new bond. Just like how Beyoncé and her dancers step together as one.

The generic exothermic reaction is a classic example of a formation reaction. Reactant A interacts with reactant B to form a product A–B. In order for that reaction to occur, a bond needs to form between A and B, which means the two species must be attracted to each other. Formation reactions can form between two atoms, two molecules, or even one atom and one molecule.

The bond formed between A and B can be ionic or covalent. Usually, this reaction is favorable because the new molecule is more stable than the original reactants. In fact, this new bond would not form in the first place if it were less stable. This interaction is similar to how a perfect couple is considered to be better together than they are apart. Because each partner brings out the best in the other, each person is happier when bonded to the other person. In a formation reaction, atoms are also better when bonded together.

The reaction between iron and oxygen is the perfect example of this. When solid iron is exposed to excess oxygen, it will rust. In this reaction, iron and oxygen go through a formation reaction to form iron oxide as shown below:

$$2Fe + \frac{3}{2}O_2 \rightarrow Fe_2O_3 + \text{energy}$$

This is an exothermic reaction, which means that iron oxide is much more stable than iron or oxygen by themselves. This is one of the reasons why rust can form so easily; iron would much rather bond to oxygen than hang out by itself.

The two basic reactions in chemistry—formation and decomposition—are relatively straightforward reactions. We either add energy to break a bond in decomposition reactions or release energy when a new bond is produced in formation reactions. But the unfortunate thing is, most chemical reactions are not that simple. Usually, multiple bonds are broken and then multiple bonds are formed. This means that enough energy needs to be provided in the reaction to break the necessary bonds in the reactants to allow for the atoms to rearrange and form new bonds in the products.

For example, let's consider two molecules A–B and C–D, where A and C are cations (+) and B and D are anions (–), in the below chemical equation:

$$A{-}B + C{-}D \rightarrow A{-}D + B{-}C$$

In order for this reaction to proceed, I need to add enough heat to my flask to break the bond between A and B *and* the bond between C and D. As soon as this happens, the atoms will rearrange to form the new bonds between A and D *and* B and C. (Remember, A and C repel each other because they are both positively charged. The same goes for B and D, but with negative charges.)

You might be wondering why A and D as well as B and C decide to form new molecules instead of going back to their original partner. The answer is simple: the new bonds are more favorable. There is a stronger attraction between A and D than there was between A and B.

Have you ever heard the story about how Ryan Reynolds and Blake Lively met? I have a point, I promise.

Ryan and Blake met on a double blind date, but they were

originally paired with other people. Ryan was with another woman and Blake with another man. Apparently, neither of them was very attracted to their original date, and basically fell in love with each other across the table as shown below:

Ryan–Woman + Blake–Man → Ryan–Blake + Woman–Man

Their awkward blind date mimics the exact way double replacement reactions typically occur. Two bonds are broken on the reactant side, and two bonds are formed on the product side. The new bonds are much stronger than the original bonds because there is now a stronger attraction between the atoms, as supported by Ryan and Blake's super cute marriage.

By the way, I'm going to be devastated if they ever break up because their story is the *perfect* analogy for double replacement reactions. For now, I'm going to assume that their relationship is solid and that they truly do have an idyllic marriage.

If Ryan and Blake represent a double replacement reaction, then Carrie and Big from *Sex and the City* are a combustion reaction. Their on-again, off-again relationship was reactive and explosive, and was generally surrounded by a lot of heat (and energy). To keep it simple, I am going to analyze one of my favorite reactions: the combustion of hydrogen. (I told you I was going to blow something up, didn't I?)

In this chemical reaction, hydrogen and oxygen gas react to form water as shown below:

$$H_2 + O_2 \rightarrow H_2O + \text{energy}$$

However, there is a major problem with this chemical equation. There are two oxygen atoms on the reactant side and only one oxygen atom on the product side. This means that the chemical equation is unbalanced, as I described earlier in this chapter. Therefore, we need to balance it by adding the coefficients to conserve the number of atoms through the reaction. The balanced chemical equation looks like this:

$$2H_2 + O_2 \rightarrow 2H_2O$$

Now, I have four hydrogen atoms on the left side (two in each hydrogen molecule) and four hydrogen atoms on the right side (two in each water molecule). I also have two oxygen atoms on the left side (two in the oxygen molecule) and two oxygen atoms on the right side (one in each water molecule).

Every time I combust a hydrogen balloon, the gas ignites, making a very loud boom. When we hear that explosion, we are hearing the result of hydrogen and oxygen atoms rearranging to form two new water molecules. Because this happens on a microscopic scale, I can never actually feel any of the water droplets that are produced.

From a microscopic perspective, this means that two moles of hydrogen and one mole of oxygen are needed to produce two moles of water (or that 1.204×10^{24} hydrogen molecules and 6.022×10^{23} oxygen molecules react to form 1.204×10^{24} water molecules). In order for this chemical reaction to occur, all of the hydrogen-hydrogen and oxygen-oxygen bonds need to break so that the new bonds between hydrogen and oxygen can be made.

To keep things simple, let's look at the chemical equation without using any coefficients. This version of the chemical equation is still accurate, but unconventional.

$$H_2 + H_2 + O_2 \rightarrow H_2O + H_2O$$

Here we can see that there are three different molecules that need to break apart in order for us to form two new molecules, but it is still difficult to see the bonding within the molecules. Therefore, we can rewrite the chemical equation as something that is more useful to us:

$$\text{H–H} + \text{H–H} + \text{O=O} \rightarrow \text{H–O–H} + \text{H–O–H}$$

In this version of the chemical equation, we have a better picture of the bonding that's happening between all of the atoms in the reaction. If we use a table of bond energies—the energy needed to make or break the bond—that's in a lot of books about chemistry (including textbooks), we can actually predict if this is going to be an exo- or endothermic reaction. The average bond energies of H–H, O=O, and H–O are 432, 495, and 467 kJ, respectively. If we plug them into our equation, we can determine if the energy change in the combustion of hydrogen is positive or negative.

$$\text{Total Energy} = \text{Bonds Broken} - \text{Bonds Formed}$$

$$\text{Total Energy} = [\text{H–H} + \text{H–H} + \text{O=O}] - [\text{H–O–H} + \text{H–O–H}]$$

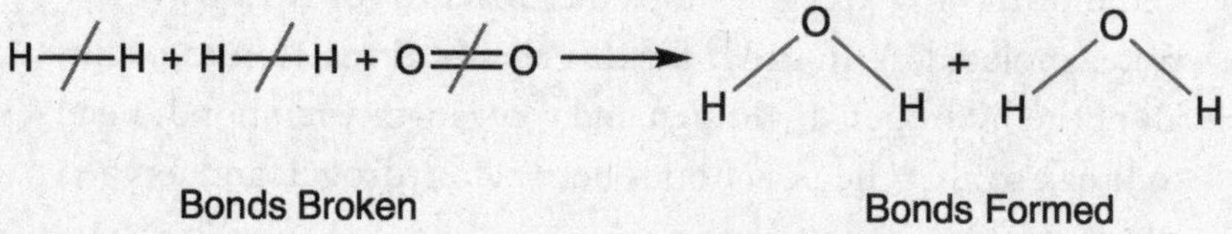

H–H + H–H + O=O ⟶ H–O–H + H–O–H

Bonds Broken   Bonds Formed

There are two equivalent hydrogen-oxygen bonds in water, so we can rewrite the Bonds Formed part of the equation like this:

$$\text{Total Energy} = [\text{H–H} + \text{H–H} + \text{O=O}] - [\text{H–O} + \text{H–O} + \text{H–O} + \text{H–O}]$$

There are two hydrogen-hydrogen bonds and four hydrogen-oxygen bonds, which means we can simplify our equation to this:

$$\text{Total Energy} = [2(\text{H–H}) + \text{O=O}] - [4(\text{H–O})]$$

Using the bond energies provided above, we can finally determine that the total energy change for the combustion of hydrogen is negative.

$$\text{Total Energy} = [2(432) + (495)] - [4(467)]$$
$$\text{Total Energy} = -509 \text{ kJ}$$

This means that this reaction is exothermic, and that the reactants are higher in energy than the products.

But what does that really tell us? The first thing this number indicates is that we would expect for this reaction to be spontaneous (i.e., to occur on its own in isolation). This should not be too surprising since most people know that hydrogen is explosive and it will readily combust.

The second thing we can learn from this number is that we should expect for the reaction to be hot. Exothermic reactions are *always* hot. They release energy in the form of heat, which we can physically feel if we are too close to the reaction.

That's important because when scientists can reliably predict the corresponding energy transfers from reactions, we can start to use the chemical reactions to create amazing technology, like hand-warmers. My husband won tons of brownie points with me recently when he remembered to bring hand-warmers on our mid-November trip to Sequoia National Park. Our early morning hikes were absolutely freezing and I was obsessed with my "chemistry hot pockets."

If you've never used hand-warmers before, you can picture small tea bags filled with a black powder. When the powder—made of iron—is exposed to the oxygen in the air, an exothermic reaction occurs, releasing heat for a few hours and keeping my hands nice and toasty. But what's freaking amazing is that the same science is used to heat small rooms in a house or to keep tropical fish warm during their transportation to their new locations. A single chemical reaction can be used for all kinds of creative applications.

Endothermic reactions, on the other hand, are cold to touch. For example, did your mom ever make you gargle with saltwater when you had a sore throat? Mine did—all the time. I would dissolve table salt into water to form a saltwater solution, and stir it for at least a minute. Then I would use the saltwater to gargle away all of the pain in my throat. But the one thing that was strange to me was that the saltwater was *always* cold. Every time I added table salt to water, the temperature of the water dropped as I stirred the salt solution. This happened every single time. Try it, and you'll see!

When most salts are added to water, an endothermic reaction occurs. The resulting solution is colder than the original liquid water, which is one commonality of all endothermic reactions.

If you've ever used an instant ice pack, then you have probably been grateful for this science at one point in your life. Instant ice packs contain two different pouches. The first pouch contains a salt, like ammonium nitrate, whereas the second pouch just contains water. Ammonium nitrate is often used in these packs because when dissolved in water, it is a particularly endothermic—or cold—reaction.

Whenever we needed medical aid on the soccer field, my trainer would grab the instant ice pack and immediately start punching it. It wasn't until later in life that I learned that you could also activate the ice pack just by squeezing it. Both methods break the two pouches inside the ice pack and force the two substances to interact. As soon as the salt hits the water, it begins to dissolve rapidly and coldly, providing instant relief to the injured player.

Hand-warmers and instant ice packs are two invaluable items

that are often found in wilderness first aid kits. It's amazing that we have been able to take two basic chemical reactions and convert them into such powerful, lifesaving materials.

Congratulations! By now you have learned *nearly* everything that I would teach in a six-week introductory course on general chemistry. You should be able to tell me about the structure of the atom, and how bonds are formed between atoms. You should be able to differentiate between ionic and covalent bonds, and explain how bonds are formed between molecules. You should be able to compare physical and chemical changes. And lastly, you should be able to explain the energy changes that occur during a chemical reaction and outline the distinctions between endothermic and exothermic processes.

And now that you have completed Chem 101, I am looking forward to beginning the second part of this book. We have established enough of a foundation of chemistry that we can discuss much more interesting topics, like the science that's in your morning breakfast, and what's *really* happening when you use shampoo. Chemistry is happening around you on a daily basis, and I think you will be amazed to learn where it is and how frequently you use it. So, go grab your apron because we are jumping straight into the kitchen!

PART TWO

CHEMISTRY HERE, THERE, AND EVERYWHERE

5

THE BEST PART OF WAKING UP

Breakfast

Now that you have mastered the fundamental principles of chemistry, I am going to take you through a typical—albeit slightly busy—day. I'll point out the science along the way and show you my favorite real-world examples. But just remember, if you need a quick refresher over any of the terms from the first few chapters, use the glossary at the back of the book to jog your memory. With that being said, let's start at the very beginning: breakfast.

Have you ever heard someone say that they're cranky in the morning until they've had their first cup of coffee? Maybe it's even you. Or maybe you've noticed that your boss is friendlier after she drinks her espresso in the morning. There is definitive evidence that coffee affects our mood, and it's largely because people easily develop a dependency on the caffeine molecule, and they feel irritable when their body is actively searching for more. Don't feel bad about it, it happens to me too. Every. Single. Morning.

Trimethylxanthine—commonly referred to as caffeine—is an odorless white powder that carries a bitter flavor. It exists naturally in coffee beans and tea leaves, therefore we rarely ever see it in its powder form. When ingested, caffeine acts as a psychoactive drug (just like nicotine or morphine), which means it messes with the way your brain functions and affects the way you behave. Some psychoactive drugs just alter your mood, but stronger ones can actually affect your consciousness. Caffeine, all things considered, is pretty mild: it just has minimal effects on our central nervous system (the part that contains your spinal cord and brain).

So how does all of that work? What happens when caffeine enters your body? How can one simple molecule give us so much "energy"? And why does it affect the way people function?

To start with, caffeine has the molecular formula $C_8H_{10}N_4O_2$ and contains the functional group purine. This means that the molecule has a five-membered ring fused to a six-membered ring, where each ring contains two nitrogen atoms. (When I say five-membered ring, it just means that there are five atoms joined together in a ring, not five atoms lined up in a row. Likewise, a six-membered ring will have six atoms joined together in a ring.)

The structure of the molecule is very important because it allows for the caffeine to bind to certain receptors in your brain. These receptors are usually trying to bond with the molecule adenosine that is naturally found in our bodies. However, the receptors get confused and accidentally bind with caffeine instead. This is a problem for the human body because adenosine is used in the production of an even bigger molecule (RNA), which is essential for human life. Luckily, the bonds formed with caffeine are short-lived so they do not permanently stop adenosine from doing its thing.

Usually, when adenosine interacts with our receptors in our brain, it makes us feel sleepy or drowsy. Therefore, when adenosine cannot bind with the brain receptors, the caffeine is able to prevent drowsiness from occurring. This means that the caffeine doesn't actually "give you energy," it really just blocks other molecules from making you feel sleepy.

Like a bouncer at a nightclub, but for your brain.

Over time, people can develop caffeinism—a condition that occurs after regularly ingesting 1–1.5 grams of caffeine a day and overtaxing the adenosine receptors. These people are usually quite easy to identify because they are often irritable and restless, and they frequently experience headaches. People can actually overdose on caffeine if they intake more than 10 grams (or 10,000 mgs) of caffeine a day. But you would have to make an extremely conscious effort to get that much caffeine into your body in twenty-four hours. It's equivalent to about fifty cups of coffee or over two hundred cans of Diet Coke.

Coffee and tea are much more potent sources of caffeine than soda. In one cup of coffee, you are likely to ingest around 100 mg of caffeine, but it can be up to 175 mg with the right coffee beans and technique. The whole process of making coffee beans (and coffee itself) is pretty fascinating if you've never given it much thought. For example, espresso makers and percolators get the most caffeine out of lighter roasted beans, but the drip method is the best way to get the most trimethylxanthine from darker beans. However, in general, light and dark roast coffees typically have the same relative number of caffeine molecules in each cup of coffee (excluding espressos).

Let's look at the roasting processes to determine why that is. When the beans are initially heated, they absorb energy in what we call an endothermic process. However, at around 175°C (347°F), the process suddenly becomes exothermic. This means that the beans have absorbed so much heat that they now radiate the heat back into the atmosphere of the roasting machine. When this happens, the settings have to be adjusted on the equipment, in order to avoid over-roasting the beans (which sometimes results in burnt-tasting coffee). Some roasters will even toggle the beans between the endothermic and exothermic reaction a couple of times, to achieve different flavors.

Over time, roasting coffee beans slowly change from green to yellow, and then to a number of different shades of brown. We refer to the darkness of the bean as its "roast," where the darker roasted coffee beans are much darker in color than the lighter roasted beans (surprise, surprise). Their color comes from the temperature at which they are roasted. Lighter beans are heated to about 200°C (392°F) and darker roasted beans to about 225–245°C (437–473°F).

But just before the beans start to, for lack of better words, lightly roast, the coffee beans go through their first "crack." This is an audible process that occurs at 196°C (385°F). During this process, the beans absorb heat and double in size. But since the water molecules evaporate out of the bean when under high temperatures, they actually decrease in mass by about 15%.

After the first crack, the coffee beans are so dry that they stop readily absorbing heat. Instead, all of the thermal energy is now used to caramelize the sugars on the outside of the coffee bean. This means that the heat is used to break the bonds in the sucrose (sugar) into much smaller (and more fragrant) molecules. The lightest roasts—like cinnamon roast and New England roast—are heated just past the first crack before being removed from the coffee roaster.

There is a second crack that occurs during the roast, but at a much higher temperature. At 224°C (435°F), the coffee beans lose their structural integrity, and the bean itself starts to collapse. When this happens, you can usually hear it by a second "pop." Dark roasts are typically categorized by any beans that have been heated past the second crack—like French and Italian roasts. In general, due to the hotter temperatures, darker beans tend to have more of their sugars caramelized, while lighter beans have less. The variation in flavor due to these methods is wild, but it doesn't really affect how they react in the body—only the taste.

Once you purchase your perfectly roasted coffee beans, you can do the rest of the chemistry at home. With an inexpensive coffee grinder, you can grind up your coffee beans to a number of different sizes, which will definitely affect the taste of your morning coffee. Small, fine grinds have a lot of surface area, which means the caffeine (and other flavors) can be extracted from the miniaturized coffee beans with ease. However, this can often result in too much caffeine being extracted, which gives the coffee a bitter taste.

On the other hand, coffee beans can be coarsely ground. In this instance, the insides of the coffee beans are not exposed to nearly the same degree as finely ground coffee beans. The resulting coffee can often taste sour—and sometimes even a little salty. But if you partner up the correct size of coffee grounds with the appropriate brewing method, you can make yourself the world's best cup of coffee.

The simplest (and easiest way) to brew coffee is to add extremely hot water to coarse coffee grounds. After they have soaked in the water for a few minutes, the liquid can be decanted from the container. This process, called decoction, uses hot water to dissolve the molecules within the coffee beans. Most current methods of coffee brewing utilize some version of decoction, which is what allows us to drink a cup of warm coffee instead of chomping on some roasted beans. However, since this method does not contain a filtration process, this version of coffee—affectionately referred to as cowboy coffee—is prone to having coffee bean floaters. For that reason, it's usually not the preferred brewing method.

By the way, did you notice that I was avoiding the term *boiling*? If you're trying to make halfway decent cup of coffee, the hot water should never actually be boiled. Instead, the ideal temperature of the water is around 96°C (205°F), which is just below boiling (100°C, 212°F). At 96°C, the molecules that provide the aroma of coffee begin to dissolve. Unfortunately, when the water is just four degrees hotter, the molecules that give coffee a bitter taste dissolve as well. That's why coffee nerds and baristas are so obsessed with their water temperature. In my house, we even use an electric kettle that allows us to select whatever temperature we want our water to be.

Depending on how strong you like your coffee to taste, you may be partial to the French press or another steeping method. Like cowboy coffee, this technique also soaks the coffee grounds in hot water, but these grounds are a little smaller (coarse versus extra coarse). After a few minutes, a plunger is used to push all of the grounds to the bottom of the device. The remaining liquid above the grounds is now perfectly clear and deliciously tasty. Since the coarse coffee grounds are used in this method, more molecules can dissolve in the coffee solution, providing us with a more intense flavor (compared to cowboy coffee).

Another technique: when hot water is dripped over coffee grounds, the water absorbs the aromatic molecules before dripping into the coffee mug. This process, appropriately called the drip method, can be done manually or with a high-tech machine, like a coffee percolator. But sometimes this technique is used with cold water, which means that the fragrant, aromatic molecules (the ones that give your coffee its distinctive smell) cannot dissolve in the water. The result is called Dutch iced coffee, a drink that is ironically favored in Japan, and takes about two hours to prepare.

One of the most popular methods, pressure, was initially used in Italy before becoming a staple of nearly every coffee shop. For this coffee brewing technique, pressure is used to force almost-boiling water through finely ground coffee. These coffee grounds have so much more surface area (compared to the coffee grinds used in the French press or cowboy coffee methods), that the water can dissolve significantly more molecules. For this reason, the resulting viscous liquid contains a *lot* of caffeine. In fact, espresso solutions contain so many molecules of trimethylxanthine (120–170 mg) that they are served in smaller cups to avoid over-caffeinating unsuspecting patrons.

My husband, like 44% of other Americans, is a big fan of coffee in the morning so he alternates between the pour-over (drip method) and the AeroPress (pressure). Personally, I'm not a big fan of the taste of coffee, therefore I was curious about what other Americans typically drink with their breakfast. Turns out, the second most popular morning beverage is water (16%), followed by juice drinkers at 14%.

Two of the healthiest juices are cranberry and tomato juice, but most Americans go with the classic OJ at breakfast. In their purest form, most fruit juices are high in antioxidants and vitamins, and low on sugar. However, their composition can be severely altered during the manufacturing process.

Let's look at orange juice as an example. If you make your own orange juice, the liquid is likely to contain a mixture of citric acid, vitamin C, and a few natural sugars. All of these molecules are soluble in the orange juice (which is mostly water), therefore they simply drain out of the fresh orange when you squeeze it over a glass.

However, if a manufacturer does the juicing process for you, they can add anything from preservatives (to prevent bacteria growth) to vitamins and minerals (like vitamin D and calcium).

Orange juice is naturally rich in vitamin C, but they add vitamin D to help promote healthy bone growth too.

In addition, the manufacturer usually performs one step called pasteurization, which is arguably more important than adding any vitamins or minerals. During this process, hot temperatures are used to break down dangerous enzymes that naturally occur in OJ. These molecules (like pectinesterase) cannot survive at high temperatures, therefore the juice is heated for about forty seconds at 92°C (198°F) before being safely packaged and shipped out to our local grocery stores.

This process of pasteurization is very common in the production of most of our fruit juices. However, the temperature and duration of time can vary greatly based on the fruit (or veggie). For my favorite juice (apple), manufacturers can heat the juice for six seconds at 71°C (160°F) or 0.3 seconds at 82°C (180°F). Since the apple is naturally quite acidic, this flash pasteurization process is all the mixture needs to prevent the growth of E. coli or Cryptosporidium parvum. And then just like orange juice, the apple juice is quickly packaged up and transported to our grocery stores.

But what if you don't drink coffee or water or juice for breakfast? What are the other common early morning beverages? According to a recent study, 11% of Americans drink soda in the morning (like me), while the last 15% drink milk and tea.

As you probably already know, milk is mostly water with some fat, proteins, and minerals. Since these fats are solid and the milk and water are liquids, this unique combination of phases is either sold as an emulsion or a colloid. An emulsion is a liquid that is mixed inside of another liquid, whereas a colloid is a solid dispersed within a liquid. In both cases, the fats and proteins have been suspended in the water to give us the thick, dense liquid properties.

Homogenized milk is a great example of an emulsion because the pulverized fats are easily suspended within the water. The little oil droplets are in liquid form and dispersed within the milk.

Conversely, whole milk is considered to be a colloid due to the high percentage of solid fat that is dispersed throughout the water. These fat droplets are even bigger, therefore whole milk cannot be emulsified like homogenized milk. However, even though these droplets are big in microscopic terms, they are still extremely difficult to detect with the naked eye.

If you are having a difficult time picturing any of this, run to your kitchen and look for a vinaigrette salad dressing. When you pull it out of the refrigerator, you will see an oil layer on top of a water layer. Now shake it. You should now be able to examine a perfect macroscopic visual of a colloid. The oil and water layers have mixed to produce an emulsion (liquid in a liquid), but all of the seeds and whatnots have been suspended into the liquid to make it a colloid (solid in a liquid).

But whether your milk is an emulsion or a colloid, if you have a little more time in the morning, you can whip up a quick breakfast that is rich in chemistry. To keep things simple, let's look at an omelet made of three items: eggs, meats, and veggies.

When you initially heat the pan, it's important for every inch of the pan's surface to warm to the same temperature. This will help us make a uniformly cooked omelet, since the atoms that make up the pan will have absorbed lots of thermal energy from the stovetop.

If we use a gas stove, the heat is being released from the combustion reaction happening directly underneath the pan. In an electric or an induction cooktop, an intricate setup of electricity and magnets are used to generate thermal energy. In most cases, copper wire is installed underneath the surface of the cooktop and then a current is passed through the wire. If you put a pan made from iron above such a cooktop, the copper wire then induces a current in the pan. Technically, any pan made from a ferromagnetic material will work; otherwise you can turn the electricity up as much as you want and your pan will never get warm. But when you use the proper cookware, the current in the iron pan triggers what's called resistive heating. This process is caused by the electrons in the metal trying to push their way through a sea of iron atoms.

To visualize resistive heating, picture a football player who is trying to run from one end zone to the other. Unfortunately for him, the football field is packed with hundreds of opposing players, and he has to do everything possible to break through all that defense. By the time the poor athlete reaches the other side of the field, he will be radiating heat from his intense workout. The exact same thing happens with electrons during induction cooking. They put so much effort into moving between the iron atoms in the pan that they release energy in the form of heat.

Since this part takes a few minutes, I like to prepare my eggs while my pan gets nice and toasty. This means that I mix two to three eggs together until I form a homogenous mixture. For this task, I prefer to use a whisk because it is gentler on the molecules within the egg membranes than other utensils. This may seem counterintuitive, but whisking is actually considered to be "gentle" on the egg membranes, which is why we use it over a fork or spoon. The whisk delicately breaks up the two sacks that make the egg whites and the egg yolk, allowing for all of the protein molecules inside the egg membranes to be mixed together. By forming IMFs (intramolecular forces) with the curvy, whisk shape, the egg whites and egg yolk can be combined without damaging the proteins.

Chemically speaking, a protein is a polypeptide—a large molecule built from two or more amino acids. There are over five hundred known amino acids, twenty of which are in our genetic code. However, only nine of them are considered to be essential. An essential amino acid cannot be synthesized by our bodies, therefore we must incorporate it in our diet.

Amino acids are commonly found in all types of food, but especially in meats. On the molecular level, all amino acid molecules have four specific functional groups connected to one central carbon atom. In chemistry, we use things called functional groups to describe small cohorts of atoms that affect the reactivity of the molecule. In other words, whenever scientists mention a functional group, we are trying to bring the focus of your attention to one small part of the molecule. It also means that the other parts of the molecule are insignificant (in the current example).

Let's focus on those four functional groups in amino acids. The first one is simple: just a hydrogen atom (H). They also have an amine (NH_2) and a carboxylic acid (COOH). No matter which of the five hundred amino acids you investigate, they will always have all three of these groups of atoms.

The fourth functional group is what gives the amino acid its identity. For example, when the fourth group is just one additional hydrogen atom, we know we are looking at glycine. But when the fourth group is an extremely long chain of hydrocarbons and an amine, we call the amino acid lysine.

Amino acids of all kinds combine to form proteins. When the amine (NH_2) from one amino acid reacts with the carboxylic acid (COOH) of a neighboring amino acid, they form a dipeptide through a carbon-nitrogen bond. The bigger proteins, like the ones found in eggs, are very large molecules, so this reaction is repeated over and over again.

When the egg proteins interact with heat, they go from a folded position to an unfolded position. Just like how a person curled up in the fetal position can "unfold" by opening their body to assume the snow angel position. The egg protein doubles in length when this happens, and with more of the atoms thus exposed, the liquid egg can then be converted into the solid egg.

The egg whites start to solidify at around 63°C (145°F), while the egg yolks change phase at about 70°C (158°F). However, when the two sets of proteins are whisked together, the IMFs formed between the molecules actually provide stability to the liquid egg mixture. This means that a whisked egg, which contains both yolk and egg white, needs to be heated to around 73°C (163°F) before it will solidify into something we can eat.

The temperature also affects the final texture and appearance of the egg. Lower temperatures will give you the white egg typical of a fried egg, while higher temperatures will give you something that looks closer to scrambled eggs. Luckily, these temperatures also take care of any of the bacteria that naturally exist within the egg.

This same statement cannot be made for all proteins, therefore it is crucial that you precook all raw meat before adding it to your omelet. The average animal protein is approximately 75% water and 25% protein, but there is a lot of variation. Each type of animal has their own unique set of proteins in their muscles, therefore the protein concentration will vary from species to species. Certain types of beef are closer to 30 or 40% protein, whereas some fish are lower at 20%. Regardless of the type of meat, they all have one thing in common: they all contain enzymes, a subcategory of proteins.

Enzymes are natural catalysts, or molecules that alter the pathway of a reaction. Catalysts typically speed up the rate at which the reactants interact because they allow a reaction to take the highway instead of the backroads. This special category of proteins played a crucial role in muscle function when the animal was alive. But now, these enzymes react within our uncooked meats and veggies, ultimately causing our food to rot. Luckily, this activity is halted when the food is stored at low temperatures, which is why we keep our food in a cold refrigerator.

However, as soon as we remove the food from the cold environment, the proteins start to react, or spoil your food. For this reason, you should leave your raw meat in the refrigerator until the moment you are ready to start cooking it. Then, add the meat to your piping hot pan to bring the temperature up quickly. There is a dangerous middle ground that you want to avoid as much as possible.

Why? Because of how the enzymes react to the heat. Just like in the egg, the heat of your pan encourages the proteins to vibrate, causing them to open up. Once the enzymes have more surface area, they can do even more damage to our food until we render them inactive.

The best way to do that is by adding heat to the meat quickly, to guarantee that you completely destroy the enzyme. Each animal protein comes with its own unique enzymes, which is why different types of meats have different minimum temperatures. For example, most beef can safely be eaten after it has reached an internal temperature of 63°C (145°F), and then allowed to rest for at least three minutes. If both of these conditions have been met, then the USDA can guarantee that the enzymes in beef (as well as any bacteria) have been killed, and more importantly, that the beef is now safe for human ingestion.

Chicken, on the other hand, must have a minimum internal temperature of 74°C (165°F) before all of the enzymes have been deactivated. Raw chicken is no joke, my friends. Both salmonella and campylobacter bacteria can be found on raw chicken, but they cannot hurt us if we take the proper precautions. For example, salmonella cannot survive being heated to 55°C (131°F) for ninety minutes or to 60°C (140°F) for twelve minutes. Therefore, the rule of thumb is to make sure the chicken has an internal temperature of 74°C (165°F). There is no way salmonella species can survive that kind of heat.

Since my husband was practically raised in a restaurant, he is super strict about where we put raw meat (especially chicken) in our household. We have cutting boards that are for raw meat and cutting boards that are for veggies only. Raw meat stays on the countertop to the right side of the stove and raw veggies remain on the left side of the stove. I'll admit that it was a little weird at first, but now I find a lot of comfort in knowing that you will never find veggies on the right or raw meat on the left in my house.

Unlike raw meat, veggies can be added to the omelet without being precooked. Just chop up whatever you're feeling that morning and throw it into the pan. My favorites are spinach, bell peppers, jalapeños, and onions. I also love to add mushrooms, but technically that's a fungus, not a vegetable.

Whereas beef can provide a hearty dose of amino acids, the building blocks of life, vegetables are essential so humans can intake a proper array of micronutrients—vitamins and minerals—that are crucial for human life.

Let's talk about vitamins first. Vitamins are big molecules that can be classified as being fat soluble or water soluble. That is such an important distinction that initially, they were all divided into these two categories. The earliest vitamins discovered were labeled A, for fat soluble, and B for water soluble. Both often contain carbon, hydrogen, and oxygen atoms, but the position of these atoms in the molecule dictates whether the vitamin can dissolve in water or fat. Vitamins that primarily contain carbon and hydrogen atoms are nonpolar, therefore they dissolve in other nonpolar molecules (fat). Vitamins that contain several oxygen atoms are often polar. That's why they can dissolve in polar molecules (water).

The vitamins in what used to be called the A group—vitamins A, D, E, and K—are all fat-soluble vitamins and can be found in spinach, mushrooms, broccoli, and kale, respectively. We used to have vitamins F–I, but it turned out that some were not really vitamins at all, just regular molecules. And over time, we learned that a few of the vitamins were actually just variations of vitamin B—like vitamin G (which became known as vitamin B_2 or riboflavin) and vitamin H (renamed as vitamin B_7 or biotin). Even though they were initially put into the wrong category, scientists were correct in identifying that both molecules are needed for basic human function.

When we eat spinach and broccoli (or any other veggies), certain vitamins dissolve in the fats in our bodies and then wait until they are needed for specific biological functions. Unlike fat-soluble vitamins, water-soluble vitamins leave our bodies through urine, which is why we can load up on vitamin C when we are sick.

For the same reason, vitamin C is a water-soluble vitamin that needs to be ingested frequently. This is a serious problem for people who do not have access to fresh produce, especially for the men and women that must work for long periods of time on submarines or on ships. Back in the early 1800s, the British Royal Navy learned that they could prevent scurvy by adding lemon or lime juice to all of the sailors' rum. Initially they didn't know why—they just observed the cause and effect—but eventually they realized the limes were providing enough vitamin C (ascorbic acid) to help the sailors avoid developing the nasty disease. Ironically, this is why the American slang *limey* was created for a British person.

For those of us that don't live on ships, we have no excuse for developing a vitamin C deficiency—or any vitamin deficiency for that matter. It's not difficult to work fruits and veggies into our diets, and even with a narrow pool of options, you'll get all of the vitamins you need.

But we don't just eat our veggies for their vitamins. Vegetables provide us with our daily mineral requirements too. Minerals are much, much smaller than vitamins because they are just charged atoms (ions), but they are all soluble in water. There are many different types of minerals that our bodies need, therefore we have sorted them into three primary categories: macrominerals, microminerals, and trace minerals.

Macrominerals are essential for human survival. Every day, you must ingest one to two grams of all of the following ions: calcium, chlorine, magnesium, phosphorus, potassium, sodium, and sulfur. Most people do this naturally by mixing up their vegetables and eating well-balanced meals. For example, broccoli has tons of calcium, lettuce and tomatoes have chlorine, and avocados contain magnesium.

When we eat, the minerals are extracted from the vegetables through the process of digestion and distributed throughout the body for basic human function. For example, calcium ions are used to produce our bones and teeth, and even to contract muscles. One of the most important functions of calcium is to help your nerves transmit signals to your brain. In other words, if you do not ingest foods that contain calcium, your body will not be able to maintain proper communication with your arms and legs.

Our bodies also require microminerals, but in much smaller quantities (hence the prefix *micro*). There are loads of different microminerals, but three common ions we need are copper, iron, and zinc. We need iron (found in animal protein) for hemoglobin to capture oxygen in the bloodstream, copper (in mushrooms or leafy greens and most nuts) to make more red blood cells, and zinc (in eggs or protein or legumes) to make genetic material for DNA.

The amount of microminerals that we need scales with the size of our bodies. For example, the USDA recommends that young children get 5 milligrams of zinc a day, whereas infants only need 3 mg. This is true for adults too, where smaller adult women only need 8 mg of zinc a day, while larger adult men need an additional 3 mg a day (11 mg total).

The last category of minerals crucial for human life are called trace minerals. We only need a few micrograms of these minerals. There are tons of different trace minerals so I'm just going to highlight the ones with the coolest jobs, like boron. Boron (in raisins) helps to manage estrogen and testosterone levels and helps maintain good bone health. Cobalt (in milk) encourages the body to absorb vitamin B_{12} and to form red blood cells. Chromium (in broccoli) breaks the fats and sugars apart, and manganese (in oysters) is needed to kick-start the chemical reactions with enzymes.

The last trace mineral that I want to call your attention to is iodine. If your diet does not contain enough of the trace mineral iodine, you can develop hypothyroidism. Back in 1990 at the World Summit for Children, a group of advocates and scientists got together to determine how to minimize iodine deficiency in certain communities. At the time, it was listed as the number one cause of preventable disabilities in children, therefore they came up with a brilliant plan to stop these unnecessary developmental and intellectual effects. They decided to replace some of the chloride ions in traditional table salt (NaCl, sodium chloride) with iodide ions (NaI, sodium iodide).

This switcheroo was incredibly successful. Just in the United States alone, we were able to prove over a few different long-term studies that the substitution of sodium iodide for sodium chloride drastically affected the overall IQ of certain communities. Not only that, it increased the average incomes for their adults by 11%. How freaking amazing is it that the concentration of a molecule in your thyroid gland can directly affect your overall intellectual ability, which ultimately contributes to your career path and salary? And how nutty is it that *salt* was the answer to this serious global health issue?

In addition to treating people with hypothyroidism, iodine can be used for those with the opposite condition, hyperthyroidism. This is a condition that occurs when someone's pituitary gland is too good at producing a hormone called the thyroid-stimulating hormone (TSH).

TSH is secreted by the pituitary gland to regulate the action of two other hormones that are produced in the thyroid gland, that help kick-start the metabolism of basically every cell in the human body. In the beginning of the mechanism, TSH leaves the pituitary gland and travels down the bloodstream until it reaches the thyroid. This is a short distance for the molecule because your thyroid gland is located near the base of your neck, and your pituitary gland is located near your nose.

Now that it's arrived in the thyroid gland, TSH encourages the release of the hormone thyroxine, followed by the hormone triiodothyronine. Both of these molecules are hormones, meaning they cause important reactions in other parts of the body (namely, metabolism). They are fascinating complex molecules which each contain the amino acid tyrosine and a bunch of iodine atoms. And of course, they are crucial for human survival.

With hyperthyroidism, the increased concentration of thyroxine and triiodothyronine can lead to any number of symptoms, including insomnia, tremors, anxiety, diarrhea, and eye bulging. It basically throws all of your systems out of whack. If there is way too much thyroxine and triiodothyronine in the body, they can cause a thyroid storm, which is a life-threatening situation caused by hyperthyroidism. When this happens, the person develops an extremely high fever (over 104°F), a racing heartbeat, and high blood pressure. This can cause a heart attack and/or liver failure, both of which can be fatal. While thyroid storms are rare, they can happen if hyperthyroidism is left untreated.

Luckily there is a solution, and it's related to our trace mineral iodine. The most common form of treatment for this condition is to take iodine in the form of radioactive iodine-131 pills. Since the thyroid cells are already predisposed to selectively bind to molecules with iodine, they actively form bonds with the radioactive iodine instead. Over time, the radioactive iodine destroys the host cells, which helps to regulate the concentration of the thyroid hormones in the thyroid gland. This means that there are fewer thyroid cells to form bonds with the thyroid hormones, which ultimately prevents the thyroid storms.

After learning about iodine in my freshman chemistry course, I began to develop a slight obsession with it. I could not get over how one trace mineral could affect our bodies in such dramatically different ways. Too much iodine (from within the thyroid hormones) can trigger a thyroid storm, which only radioactive iodine can fix. And too little iodine contributes to brain degradation. It's a really hard balance to manage, which is why making smart food choices is very important.

So, from caffeine to cooking eggs to choosing the right fruits and veggies, we've already gotten our day off to a great start. In the next section, I am going to evaluate what happens after we eat our breakfast. I'll look at how our bodies process the food and convert it into energy that can then be put to work at the gym.

Fair warning, I used to be a fitness instructor and I tend to get pretty jazzed about exercise. In fact, hold on one second. I need to go grab my aerobics microphone that helps me channel my inner Jane Fonda.

6

FEEL THE BURN

Working Out

I'm an adrenaline junkie.

I love things that are loud, fast, and just a little bit unsafe. But since I can't go skydiving before my 8:00 a.m. chem lecture, I usually just get my kicks from an early morning workout.

It's my grandpa's fault really. He used to jog every morning "before it was cool" (according to him). And for as long as I can remember, my dad has worked out every single morning—even on vacation. In fact, there is so much garage sale workout equipment in my parents' basement that you would think my family runs some kind of underground exercise club.

And now, even though I have fought it off for as long as possible, I have started to work out every other morning. My dad is convinced that I will graduate to one-a-days when I get older, and he might be right. I really do enjoy exercising.

When I was in graduate school, I started searching for fun things that I could do outside of the lab, and somehow I decided to become a fitness instructor. For a few years, I taught step classes and led early morning boot camps, but my favorite class to teach was Turbo Kick (i.e., choreographed kickboxing).

Toward the end of my tenure as a fitness instructor, I was hired as a Nike Training Club instructor. This was an invaluable opportunity for me at the time because I was dirt broke and Nike gifted us a duffel bag full of gear every fall and spring semester. In return, we had to wear the attire (no objections here) and attend a few information sessions. While my fitness buddies were less inclined to attend the seminars, I was stoked to hear the science behind all of their workout gear.

Some of the information given to us was relatively obvious, like how running shoes are designed to have a cushier sole than cross-training shoes. The runner is elevated higher off the ground, which is why running shoes are not intended to be used for lateral movement. Running shoes are strictly designed to help minimize the force applied to the joints of the runner, whereas cross-training shoes give you lateral stability because the athlete is lower to the ground.

My ears perked up when they started analyzing their unique Dri-FIT material. Moisture-wicking fabrics, commonly referred to as Dri-FIT, are used for a type of clothing that has been engineered to help athletes moderate their body temperature while exercising. Some fabrics, like cotton, are notoriously bad at this, which is why you may have heard of the phrase "cotton kills." But why is that?

In general, the human body cools itself by sweating. Water molecules are pushed out through our pores, and then they bead into droplets on the surface of our skin. At this point, the most important part of the process occurs: the water molecules evaporate. When the water transitions from the liquid to the gas state, it pulls heat from the closest source of energy (your body), which ultimately cools your overall body temperature.

With this information, Nike decided to use a blended polyester that can pull the water droplets off the athlete's body. Once the molecules have been absorbed by the special threading in the fabric, the water molecules can glide across the material. This process exposes more water molecules to the heat in your body, which allows for evaporation to happen more quickly. Ultimately, the more water molecules that are absorbed by the fabric, the more heat that is pulled from your body, which drops your overall body temperature at a faster rate.

On the other hand, cotton fabrics have the opposite effect. The threads are woven together so tightly that the water molecules cannot easily evaporate into the atmosphere. They are physically trapped between the skin and the fabric, forcing them to remain in the liquid state for a much longer period of time.

Nowadays, since Nike is no longer supplying my workout gear, I usually order attire that is made from a mixture of polyester, nylon, and/or spandex. These three fabrics are breathable and have wide enough gaps between their threading that water molecules can evaporate from the athlete's body.

I tend to vary my morning workout attire based on my mood. When I'm grumpy, I wear my robot pants from INKnBURN; when I'm happy, I might go with my hot pink pants from Amazon. Regardless, after I've changed into my wicking material, I grab my reusable water bottle and stop by the kitchen.

All foods, regardless of their composition have Calories. But I'm not talking about what you see on a nutrition label. A Calorie (uppercase *C*) is a *nutritional* unit of energy and the calorie (lowercase *c*) is a *scientific* unit of energy. One Calorie is equal to 1000 calories, so 1 Cal = 1000 cal = 1 kcal. We use nutritional Calories on nutrition labels because it is much easier to say that there are 140 Calories in eight peanut butter pretzels than 140,000 calories.

But what does that number really mean? There are several different ways to think about it, but to me, 140 Calories means that our bodies can convert eight pretzels into 140,000 calories of energy. In other words, that would give us enough energy to stretch for about an hour or maybe even go for the slowest walk on a treadmill.

Eight pretzels can be converted into enough energy for us to move our bodies around slowly for an hour, while sixteen pretzels would give us enough energy to go for a very light bike ride. Pretty amazing, right? Especially considering that we do this without even thinking about it.

The conversion of food into energy is a long and arduous process referred to as oxidative phosphorylation. It occurs in the presence of oxygen and can be simplified into three easy steps.

The first step is not surprising—your stomach has to digest the food. The enzymes in your stomach and large intestine attack the molecules in your food to break them down into much smaller units. When the food molecules are really big, the process takes a very long time.

Once all of the big molecules have been separated into the much smaller molecules, glycolysis occurs, the second step in this process. During glycolysis, glucose is broken in half to form smaller molecules called pyruvates. These pyruvates are then converted into carbon dioxide (which we exhale) and two other molecules that contain acetyl functional groups (H_3CCO). Both of these new molecules bind to coenzyme A to produce acetyl CoA before being transferred to another molecule called oxaloacetate.

When this happens, the acetyl molecules enter the citric acid cycle before being transformed into carbon dioxide (which we again exhale). This process generates NADH molecules, which kick-starts a process that produces ATP.

And that, my friends, is the entire point of this digestion process: to make ATP.

ATP—adenosine triphosphate—is one of the most important molecules in our body because it provides energy to our cells.

This energy helps our nerves send signals to our brain and even helps muscles to contract. It is so important that ATP has been described as the "molecular unit of currency." But as you may have realized, ATP cannot remain in the stomach or large intestine. So, where does it go?

The energy is stored as small packets of energy within the cells all over the human body. With this technique, our bodies are able to release energy in an instant whenever we need it, wherever we need it. For example, if you need to run to catch a bus or if you instinctively reach for something that has been knocked off a table, your body releases energy fast enough to let you move through your day-to-day activities *quickly*—and without going into hibernation immediately afterward.

The number of instant movements that we can make throughout our day are directly proportional to the amount of ATP that we have stored within bodies. At any given moment, you can expect to find approximately one billion molecules of ATP in any cell anywhere in your body. One *billion* molecules in *any* cell in your body.

And then two minutes go by.

And in that time, *all* of those ATP molecules have been used up *and* regenerated.

Think about that for a second. One billion energy molecules are moving around in each of your cells in your body right now. But in two minutes, they will all be converted into other molecules (like ADP and AMP), and then back into ATP. Then back to ADP or AMP, and then back to ATP. This process never stops until the cell dies, which can happen for any number of different reasons.

So, if we bring this back to our pretzel example, we said that eight pretzels give us 140,000 calories (140 Calories). But how does that relate to ATP? Well, after we eat the pretzels, our bodies break down the food into about nineteen moles of

ATP, which is then burned to release 140,000 calories of energy (1 mol of ATP = 7.3 kcal).

And as we discussed earlier, 140,000 calories is barely enough energy for me to go for a walk. In fact, as an active woman in my midthirties, I am supposed to intake approximately 2200 Calories a day. Do you know how many of those calories go toward just keeping me alive?

Guess.

Did you say 1300? Or just under 60%? If you did, then you are right. I need 1300 Calories to continue living on planet Earth. This is the minimum amount of energy that I need daily to keep my heart beating, my lungs pumping, my brain thinking, and my core body temperature at 37°C (or 98.6°F). With fewer than 1300 Calories, my body will naturally start to prioritize where the energy is spent. I will begin to feel exhausted, which is my body's way of telling me to take a nap, because my body is using backup energy to make sure that my internal organs do not shut down. In theory, my body could last about three weeks on reserve energy before any organs began to malfunction.

If 1300 Calories go toward just keeping me alive, that means that the extra 900 Calories can be used for my morning workout! For example, I need about 500 Calories for an hour of swimming (or heavy yard work) and about 330 Calories for an hour of Zumba.

In general, most active people burn through the remaining 40% of their ingested Calories every day. However, people that live a sedentary lifestyle have extra Calories remaining at the end of the day. In this case, these Calories are stored as fat to be used in case of an emergency. If you think about it, your body is trying to be helpful by keeping extra energy around in case you are unable to eat your full 2200 tomorrow. If you regularly eat 2200 Calories (or more), your body will have no reason to dip into its fat storage tank for energy. This, of course, can eventually lead to obesity.

But what actually happens when you hit the gym in the morning? In the beginning, our bodies reach for the ATP stored in our fat cells before searching for any carbs or proteins. This is because 1 gram of fat releases 9 Calories of energy, whereas 1 gram of carbs or proteins release only 4 Calories of energy. Fat is simply a better source of fuel for our body.

Why is that? Fats are actually stored in something called the adipocytes, which is a cell whose sole function is storing fats. Just like glycolysis separates glucose into pyruvates, lipolysis separates lipids (fats) into three fatty acids and one glycerol molecule. Once the lipids have been broken down, the fatty acids leave the adipocyte cell and enter the bloodstream. Here, they are carried by the albumin protein to the muscle cells, where they can move through our capillaries directly into the muscle.

When we exercise, these proteins appear on the outside of the muscle membrane before the fatty acids are converted to ATP—just like how the glucose molecules were converted to ATP. In both processes, heat is needed to break the covalent bonds in the fatty acids and glucose molecules, before ultimately producing ATP. This entire process is referred to as aerobic metabolism.

The word *aerobic* means with oxygen, and that's why we call the classes at the gym aerobics classes, because the process of burning fat occurs in the presence of oxygen. Have you ever noticed that an intense workout is associated with heavy breathing? This is because we are desperately sucking in oxygen in order to burn enough ATP to give us enough energy to complete the workout.

With this in mind, it is probably not too surprising to see that the harder you work, the more oxygen you inhale, and your oxygen consumption directly correlates to the amount of fat/carbs that you burn. For example, at 25–60% oxygen consumption, the oxygen is used to burn the fat that is located in your bloodstream—this is the fat from the food that you ate. However, if you start consuming 60–70% oxygen, your body switches from burning the fat in your bloodstream and now begins to use the fat in your muscles. Above 70%, your body kind of panics and starts to use carbs as source of fuel.

This part never made sense to me in high school biology. Why does our body switch from fats to carbs when our consumption of oxygen is above 70%? It's not like all of the fat has been used up. We clearly still have fat on our bodies—in our booties—so why does our body act like we have run out?

Well, it has to do with the location of the fat. When we are in high intensity workouts, our muscles do not get as much blood, therefore they run out of fatty acids to burn for energy. In fact, the body actually directs the blood away from the fat tissue. This indicates that the fatty acids are still released into the capillary tubes, but they are stuck there. They can no longer move through the cell membrane into the muscle cell, therefore they cannot be used as a source of energy during intense trainings.

It's almost like if all of your fatty acids were stuck in your backyard. You can see them through the window, you know they are there, but you cannot access them until the backdoor has been opened. So, in the meantime, you run to your pantry to find a backup source of fuel: carbohydrates. Your body does what it can with the low energy fuel source, until it can inevitably switch back to burning fats.

The cool part about a good workout is that our bodies continue to burn fat even after the workout has ended. It is referred to as excess post-exercise oxygen consumption or EPOC, and HIIT classes (high intensity interval training) are notorious for having high EPOC. Why? Because classes like CrossFit or Nike Training Club destroy the muscle tissues so much during the workout that your body has to work overtime to repair all of the damaged muscle cells. In order to bring them back to pre-workout levels, your body needs to replace the glycogen in the muscles in addition to fixing all of the damaged cells.

Now, I want to address the fact that I am clearly biased here. I used to teach fitness classes and to this day participate in interval trainings sessions weekly. I prefer HIIT over other endurance trainings—like running or biking—because I have had four ACL knee surgeries. I can no longer do anything high impact, like running, and I get too distracted to stay on a bike for a long period of time. However, both running and biking are great for your heart and your cholesterol levels because they burn high levels of fatty acids *during* the workout session. They also encourage the burning of fats after the workout, but not nearly to the degree of HIIT classes.

Regardless of exercise, have you ever noticed that every workout program gets easier over time? This is because you are starting to train your muscles. In other words, you are starting to teach them how to operate and function properly. The adipocyte released is going to be the same for sedentary people and super active people. The difference is that a trained muscle can easily uptake the fatty acids and quickly convert them into energy. This is because strong athletes have a higher mitochondrion count per unit of muscle and ATP burns *inside* of the mitochondria. The more mitochondria, the more fat burned.

So now, here is the million-dollar question: How do you actually lose weight? Where does it go after it has been burned? If you have been paying attention, you may have picked up on the fact that every time we burn ATP, we release carbon dioxide. This means that all the fat, proteins, and carbs that you burn during a workout are released from your body through your exhale.

Can you believe that? You exhale your fat. That's how you lose weight. It does not happen when you go to the bathroom or when you sweat, it literally leaves your body through the molecules that you breathe out of your mouth during (and after) your workout.

And even though working out can give you a toned body and a healthy heart, the best part of any workout for me—like I mentioned earlier—is the adrenaline rush that comes with the strenuous exercise. Epinephrine, commonly referred to as adrenaline, is an amino acid-derived hormone that has the molecular formula $C_9H_{13}NO_3$. It has one six-membered ring and three alcohol functional groups across the molecule, which give it its unique physical properties inside the body. In humans, the molecule is secreted from the adrenal glands, which are located directly above the kidneys.

What's particularly neat about adrenaline is that when it is released into the bloodstream, it can interact with our body tissues, but it has different effects on different organs. For example, adrenaline generally increases your respiratory rate and causes vasodilation (the opening of the blood vessels). Yet somehow, in other parts of the body, it can cause vasoconstriction (the closing of blood vessels) and muscle contractions.

Because of these properties, adrenaline can be used as a lifesaving medication. For example, if someone suddenly experiences a severe allergic reaction to something, we can use an epinephrine autoinjector—or an EpiPen—to quickly inject adrenaline into the outer thigh of a person going into anaphylactic shock. A solution of epinephrine goes into the muscle and is quickly absorbed into the bloodstream. The medication then causes vasoconstriction (in order to increase blood pressure) and opening of the lungs, which allows for the poor person to start breathing again.

These adrenaline rushes can happen naturally in the gym too. After an intense workout, our bodies release epinephrine, but they also release a hormone called dopamine. This molecule acts like a reward for the body when any goal (like an hour of rigorous exercise) has been accomplished.

Dopamine molecules look very similar to adrenaline molecules, but they only have two alcohol groups (OH). They are water-soluble molecules so they easily move through your body to reach dopamine receptors. When that happens, our body is overcome with a euphoric feeling. For certain people (like me), this feeling can be extremely addicting. It can be so powerful that it can cause reward-motivated behaviors, like an athlete training harder in the gym or a dancer putting in longer hours at the studio. Just knowing that the reward is coming can be enough to trigger the release of dopamine. For that reason, scientists are hesitant to say that adrenaline junkies are really people seeking "adrenaline." Instead, they believe that certain people are prone to taking big risks without any regard for physical or social safety because their bodies want the reward of dopamine. We should really call these people "dopamine junkies."

Adrenaline is such a powerful molecule that it has been attributed to giving humans superpower-like strength, called hysterical strength. In 2019, a 16-year-old football player from Ohio heard his neighbor call for help. Upon learning that his neighbor's husband was trapped beneath a 3,000-pound car, he jumped into action and was able to generate enough strength to lift the car so that the man could be moved to safety—all because of a surge of adrenaline.

For that same reason, some athletes use performance-enhancing drugs that have been spiked with adrenaline. This is because adrenaline increases stamina and endurance and also provides increased physical strength. In certain sports, it lends itself to giving reduced reaction times, which gives the cheaters a serious edge over their competitors.

But adrenaline is rarely found alone in the body. In addition, the body produces another hormone called cortisol. This molecule causes both your blood pressure and your blood sugar to increase, in addition to getting ready to turn your fats into energy. The muscles essentially feel the presence of cortisol and prepare themselves for quick, strong actions, like a burpee or a squat jump.

Cortisol is a steroid hormone, which has the typical steroid structure of four carbon rings fused together. On one side of the molecule there is a ketone (C=O) attached to the rings, and on the other side, there are a few alcohol groups (OH). Since the oxygen atoms are equally spaced out over the molecule, cortisol is relatively nonpolar. Therefore, it is more likely to form dispersion forces with any neighboring molecules.

This hormone can do a number of different things in your body. Depending on where it is, cortisol can contribute to how that part of your body operates. For example, it can increase the concentration of your blood sugar or alter the way your metabolism functions. Unlike adrenaline and other amino acid hormones, steroid hormones are fat soluble (not water soluble) because they are nonpolar molecules.

The hormone cortisol plays a crucial role in gluconeogenesis, which is the process of forming the glucose molecule from nonsugars. As we discussed earlier, glucose is an excellent source of energy for the body. If our bodies are "out" of sugar, then we can use gluconeogenesis to force a chemical conversion of fats or proteins into glucose.

For this reason, adrenaline and cortisol are the two primary hormones that are released by the body in response to stressors. It doesn't matter if you are doing a morning kickboxing class or running away from some bad guys, your body has the same physiological response.

But remember how I mentioned that adrenaline is water soluble and cortisol is fat soluble? This is because they have different polarities. Adrenaline is a polar molecule, which means it can use your bloodstream like a river to float to its target organ. Since cortisol is a nonpolar molecule, it needs to use carrier proteins, like floaty devices, to move to fat cells, where they can begin to be effective.

When I talk about these hormones in class, my student athletes always follow up with a logical question: Are adrenaline and cortisol also responsible for the runner's high? I have to begrudgingly answer with the typical scientist answer "yes, but..." because the human body has so many different variables. For example, when your body is under any type of stress, like a cross-country race or a football game, it will definitely release adrenaline and cortisol to varying degrees. But at the same time, it releases endorphins.

Endorphins were first isolated in the 1960s when biochemist Choh Hao Li was studying the dried pituitaries of five hundred different camels. He was looking for a specific molecule that metabolized fat, but he could not seem to find it in the camels. Instead, he discovered a polypeptide, which later became known as the beta-endorphin. But he had no use for it at the time, so he carefully wrapped it up and tucked it away for safekeeping.

About fifteen years later, Li heard about research performed by biochemist Hans Kosterlitz and neuroscientist John Hughes. They had discovered a pentapeptide—or a molecule with five linked amino acids—called enkephalin. After hearing about these results, Li pulled out his beta-endorphin to see if it too had enkephalin. It did, so he decided to test it as a pain reliever in comparison to oxycodone or heroin. When he injected it into the brain, he was ecstatic to observe that depending on the injection site, it was anywhere between eighteen and thirty-three times more powerful than traditional morphine. Unfortunately, it turned out that enkephalin was *significantly* more addictive than morphine, so he abandoned the idea of converting it into medicine.

Over time, scientists began calling enkephalin and any endogenous neuropeptides by the generic term *endorphins*. In today's vernacular, if a molecule had analgesic properties and caused euphoria, then the substance could be called an endorphin.

In general, humans produce three different types of endorphins in our bodies naturally: α-endorphin (alpha), β-endorphin (beta), and γ-endorphin (gamma). The α-endorphin molecule is a chain made from sixteen linked amino acids. The γ-endorphin molecule is nearly identical; it just has an additional leucine amino acid at the end of the chain.

The β-endorphin is a much bigger molecule made from thirty-one connected amino acids. Just like before, the first sixteen amino acids are identical to those of the α- and γ-endorphins. The last fifteen amino acids are a mixture of leucine, phenylalanine, lysin, and glutamic acid, just to name a few. But unlike the other endorphins, the β-endorphin has been shown to have a number of different effects on the human body. For instance, research suggests that it plays a role in reducing stress when people are in pain or in hunger. It also activates the reward system and some sexual behaviors.

In the 1980s, scientists were able to link the β-endorphin to what has become known as the runner's high. They learned that upon intense exercise, the body releases β-endorphins, which helps us manage pain. Usually when we experience pain from an intense workout, pain receptors use a molecule called substance P to send signals to the brain through our spinal cords. However, when this happens, the body sends β-endorphins out too—like medics on a battlefield—to help moderate the pain. The endorphins form a bond with the opioid receptors in our spinal cord, blocking substance P from binding to the same receptors. This chemical reaction is a crucial step in minimizing pain.

When our workout is really intense, like after doing sprints or maxing out on weights, there can be a high concentration of endorphins bound to the opioid receptors in our brains. This brain chemistry gives us that amazing, euphoric feeling that we experience immediately after we stop the activity. When a person is in a particularly stressful moment in their lives, this release of endorphins can push people to have a strong emotional release. Just look at a video of gymnast Aly Raisman as she finishes her floor routine at the 2012 Olympics. She bursts into tears as she lands her last move, knowing full well that she just became the first American woman to secure a gold medal in the floor exercise.

In 2012, I would have explained her reaction as a high concentration of endorphins connected to opioid receptors. But then in 2015, a group of German scientists realized that endorphins cannot pass through the blood-brain barrier. Even though endorphins reduce anxiety and minimize pain, they actually cannot provide the high felt after a strenuous workout. Intrigued by this result, they performed more experiments to learn that a molecule called anandamide had no problem moving from the blood to the brain to trigger the mood elevation.

So, what's anandamide?

Anandamide is a fatty acid molecule that binds to cannabinoid receptors, which are the same receptors that form bonds with cannabis (weed). The name aptly comes from the prefix *ananda*, which means bliss and joy. Because it is responsible for how we experience pleasure, this molecule is sometimes referred to as the bliss molecule. Interestingly, this molecule was only discovered because of the recent research on marijuana. Researchers wanted to know more about how weed (tetrahydrocannabinol, THC) operates within the human body, and learned all about the cannabinoid receptors.

It turns out that cannabis interacts with cannabinoid receptors just like how opioids bind to opioid receptors. However, there is one major difference. The opioid-opioid receptor bond is extremely strong, whereas the cannabis-cannabinoid receptor bond is relatively weak. The weed bonds break within a short period of time, which doesn't allow the cannabinoid receptors to build up a dependency on them. This weak interaction is the reason why weed is not nearly addictive as OxyContin and heroin.

But what does that have to do with working out? Well, the weak anandamide-cannabinoid receptor bonds are one of the reasons why the runner's high does not last very long. It is simply not strong enough, therefore it eventually breaks to lessen the runner's high. And after your runner's high is gone, you may notice one thing: pain.

Since I have had four ACL knee surgeries, my most common source of post-exercise pain is in my knees. Over the years, I have eliminated all forms of jumping in my workouts, but sometimes, my knees just ache. In those instances, I do not hesitate to grab some over-the-counter pain meds.

But what exactly is a pain medicine? And how do these molecules work inside of our bodies?

Aspirin, the "wonder drug," as it has been called in more recent media, was actually first reported by Hippocrates back in the fourth century BC. At the time, they boiled willow bark in water to make a tea that cured fevers. Fast forward to 1763, when an English clergyman named Edward Stone published a letter that revealed new research on willow bark. He reported that after drying willow bark, he gave portions to fifty different people. Those people used the dried pieces of bark to cure common ailments, or even as a weird form of lotion.

That said, Stone's patients who ingested the tree bark all complained of the same two problems: (1) the willow bark had a nasty taste, and (2) it gave them bad stomach problems. Yet his patients were willing to keep eating the bark because it lessened the pain from headaches and inflammation. And in some cases, the bark even relieved the symptoms of arthritis.

Over a hundred years later, a chemist named Felix Hoffmann started pursuing a chemical alternative to salicylic acid ($C_7H_6O_3$), the active molecule in willow bark. His father was experiencing severe nausea from his salicylic acid meds, and Felix wanted to see if he could help him out. With the help of his boss, Arthur Eichengrün, Felix began tinkering with salicylic acid and eventually discovered an efficient way to make its close cousin: acetylsalicylic acid ($C_9H_8O_4$), later called aspirin.

Unfortunately, Arthur and Felix had great difficulties getting any clinical trials of the new drug because salicylic acid was notorious for weakening the heart. Felix was reassigned, and used his new knowledge of acetylsalicylic acid to synthesize another popular drug called diacetylmorphine—commonly known as heroin. Surprise, surprise, he had no problem convincing people to try diacetylmorphine.

Since Arthur was the boss, he was not going to give up on acetylsalicylic acid that easily. He sneaked the new pain medicine (what we now know as aspirin) to doctors, who started their own private clinical trials. The results came in immediately because the patients (and their doctors) were ecstatic to finally have something that cured their fevers and relieved their pain without giving them severe stomach discomfort. The news spread quickly, and within a short period of time, Bayer Aspirin was available over the counter.

So, what was so great about aspirin (acetylsalicylic acid) and why did it seem to be better than willow bark (salicylic acid)? Initially, chemists at the time realized that the replacement of one alcohol group (OH) in salicylic acid with an ester group ($OCOCH_3$) in acetylsalicylic acid made the aspirin taste better and alleviated the stomach problems. They also learned that the drug was equally as active as salicylic acid. But how could this be? The bigger molecule ought to have had a harder time reaching the desired location simply due to its size.

What scientists quickly figured out was that Arthur and Felix hadn't actually discovered a new drug. Aspirin (acetylsalicylic acid) may have been easier for people to ingest, but in the stomach, it decomposed back into salicylic acid. Essentially, aspirin just lessened the amount of negative side effects that willow bark had on the stomach and mouth.

Aspirin is great to take after a sports injury because it helps to minimize the pain we feel from inflammation or swelling. Inside the body, the salicylic acid blocks an important chemical reaction so that the enzyme (cyclooxygenase) can no longer produce two molecules (prostaglandins or thromboxanes). Prostaglandins cause vasodilation (or opening of the blood vessels), which sends white blood cells to the injury. In other words, the aspirin stops an enzyme from making our ankle swell up after a bad fall.

This process is also very common in the mechanism for NSAID drugs—or nonsteroidal anti-inflammatory drugs. Ibuprofen, for example, is another NSAID that is commonly used to treat inflammation, pain, and fevers. It is a much, much newer drug compared to aspirin. Discovered in the 1960s, ibuprofen was also found to inhibit the activity of cyclooxygenase. This molecule has the molecular formula $C_{13}H_{18}O_2$ with one six-membered ring in the middle of two hydrocarbon chains. As you may already know, it works extremely well to treat fevers and can help to minimize the pain caused by kidney stones.

Acetaminophen, an even cheaper drug, is commonly called Tylenol. Chemists usually refer to it as paracetamol, and it can relieve some of the symptoms of the common cold. This drug has the molecular formula $C_8H_9NO_2$ and contains a six-membered ring.

The science behind how acetaminophen operates is still not

quite perfectly understood. Unlike ibuprofen and aspirin (and other NSAIDs), acetaminophen does not seem to block the cyclooxygenase enzyme. Instead, it appears to function through an entirely different pathway. It is still believed that the enzyme is involved, but researchers are not 100% sure how that process actually works. The one thing they know for sure is that since it does *not* block the enzyme, acetaminophen does *not* work as well as an anti-inflammatory drug. However, scientists are starting to believe that it might be able to inhibit the enzyme in the brain, which could be a reason why it can be used to relieve pain and treat fevers.

Each molecule has different effects on the human body, which is why certain pain medicines seem to work better for different injuries. Personally, I'm partial to Aleve (naproxen), which is another NSAID. Since I am not a medical doctor, I am not going to recommend one pain medicine over the other, but I will caution you to pay attention to any negative side effects, like organ damage. For example, I have to be very careful not to overuse Aleve because it can cause kidney failure.

But no matter how much pain I feel in my knees, it will never keep me out of the gym (I hope). I love starting my day with a runner's high and wearing cute workout gear. And even though I shouldn't, I use it as an excuse to give into my sweet tooth.

On most mornings, a good workout wakes me up and gets me ready to attack the day. However, before I can go out in public, I need to take a few minutes (or an hour) to get cleaned up. In the next section, I will break down the science that surrounds you in the bathroom. From your shampoo to your blow-dryer to that hot red lipstick you've got the confidence to wear to work, they all have one thing in common. You guessed it: chemistry!

7

BE·YOU·TIFUL

Getting Dressed

After an intense workout, I usually head to my bathroom to begin the process of getting ready. For the longest time, it was my least favorite part of the day. In grad school, I never had enough time to do much more than toss my hair into a wet ponytail and hastily throw on a T-shirt before running out the door. Now, I have a whole morning routine and a makeup collection that admittedly takes up more than my share of bathroom space. Not surprisingly, from hair care to makeup to perfume, it's all chemistry—and in the right combination it can have nearly magical results.

Believe it or not, there is extraordinary science even in a nice, peaceful shower. When the hot water rains down on your body, the water molecules are forming hydrogen bonds with the neighboring water molecules on your hair and skin before they even reach your shower floor. Sometimes, their adhesive properties are so strong that they pull the water molecules away from your epidermis to form water droplets on your skin. When this happens, the water molecules are more attracted to each other than the molecules—or salt on your skin.

The science of our shampoos and conditioners is even more interesting. Most of them contain molecules you probably haven't heard of (and won't see listed on the back of the bottle): quats and cationic surfactants. They're not as complicated as they sound, I promise—and their interactions with hair are irreplaceable. But before we get into those unique compounds, we have to start with the basics.

The primary protein in hair is called alpha-keratin. You may have heard of keratin before, in the context of nail growth and skin care, or even in horns and feathers on animals. Your salon may even provide keratin hair treatments to temporarily straighten curly/wavy hair, and it could be one of the additives in your shampoo or conditioner.

Keratin, whether in our hair or in a bottle, is made from a polypeptide chain of amino acids. The order, arrangement, and identity of the amino acids will vary, but the overall protein (alpha-keratin) will always contain at least one cysteine molecule. This relatively small amino acid can sometimes act like an enzyme and trigger biochemical reactions. How? When two polypeptide strands (or two keratin strands) wrap around each other to form a coiled coil (yes, that is really what it is called!), the sulfur atom in the cysteine molecule from keratin A forms a covalent bond with the sulfur atom in the cysteine molecule from keratin B. This reaction produces a completely new molecule called cystine. And although their names are almost identical, cystine is the bigger molecule formed by the combination of two cysteine molecules.

What you need to know is that this process, repeated over and over, forms a ladderlike structure (like DNA), where each rung of the ladder represents a sulfur-sulfur bond between cysteine amino acids. This chemical reaction is extremely important because the resulting cystine molecule (each rung of the ladder) is where all of the chemistry happens!

Any time you wash, dry, or straighten your hair, you are messing with your cystine molecules. Since we usually start our shower by washing our hair, let's look at that process first. Shampoo removes the grease and oils on our head, which is heavy-duty cleaning, yet it cleans without causing the skin on our scalp to burn or sting. Scientists have selected chemicals that can form bonds with the offending molecules in our hair but are gentle on the hair itself, and can still be safely and easily flushed down the shower drain.

Each brand of shower products has their own unique combination of molecules that work together to remove the lipids, bacteria, and unwanted byproducts from our hair. These molecules can be viscous liquids (like the super thick solvent glycol) to citric acid (found in lemons) to salts (like ammonium chloride). But my guess is you have heard more about parabens, sulfates, and silicones in the context of shampoos and conditioners since these molecules are frequently in the headlines.

Let's start with parabens, since they are used to block bacteria growth in many cosmetic products. The most common parabens are *meth*ylparaben, *eth*ylparaben, *prop*ylparaben, and *but*ylparaben, all of which are derived from parahydroxybenzoic acid. In the language of chemistry, meth = 1, eth = 2, prop = 3, and but = 4. These prefixes indicate the number of carbon atoms within each paraben molecule.

Methylparaben is a very common antifungal preservative that is added to our shampoos, in addition to many of our food products (*antifungal* meaning it will break the bonds within fungi and bacteria, which can't easily replicate or survive in its presence). If you live in Europe, you can easily identify methylparaben by the E number, E218. Ethylparaben (E214) and propylparaben

(E216) are also used in shampoos and conditioners, but they are not nearly as popular for antibacterial purposes as butylparaben.

Butylparaben has a four carbon atom chain attached to its main oxygen atom in the paraben molecule. Because of its longer hydrocarbon chain, it has different physical properties than its molecular paraben cousins. Butylparaben is used in over 20,000 cosmetic products, but most people are surprised to learn that it is also found in common medications, like ibuprofen.

Unfortunately, while they do a great job preventing bacterial growth in our shampoos and conditioners, parabens have been connected to a few adverse human health effects. In 2004, a study was released stating that parabens were found in the tumors of eighteen out of twenty women with breast cancer. At the time of the publication of this book, no other studies have been published with more information. Does it mean, definitively, that parabens increase our chances of getting breast cancer? Not necessarily. But many women—including this one—try to avoid it in our cosmetic products all the same, and companies have responded by making a wide variety of products paraben-free.

Many manufacturers use paraben-free products by taking advantage of airless apparatuses to dispense their products. Airless apparatuses are exactly what they sound like: bottles that are sealed after all of the air has been removed from the space above the shampoo or conditioner. As you may imagine, this is a feat that is exceptionally difficult to do in the shower. However, it's an important step because the lack of oxygen makes it difficult for bacteria or fungi to grow.

Another molecule that I try to avoid is silicone. Not because it might cause cancer, but because it is the main molecule responsible for what is referred to in the hair industry as *buildup* (a coating that makes your hair feel thick and slick). Unfortunately, it took way too long for me to learn that the greasy flat hair I rocked in middle school was due to the silicone in my shampoo, and that showering with the same product every day was *not* the answer.

Polysiloxane, commonly referred to as silicone, is a large polymer that is probably already in your shower or bathtub—in the caulk. Since it's a nonpolar molecule, silicone is water resistant with serious thermal stability. In shampoos, it can coat each hair follicle and temporarily grant the hair strands these same properties to protect it from environmental damage. It also reduces frizz; if each strand of hair has a soft "rubbery" texture, then the hairs should be able to glide right past one another.

However, like I said before, there is one major side effect of the silicone coating: it can cause major buildup over time. Because it is a heavy material, it clings to the hair and weighs each strand down, and over time, the silicone forms adhesive bonds with other silicone molecules, giving some users the greasy hair look I mentioned previously.

Another vilified shampoo ingredient, at least to some minds, is a surfactant called sulfate. Sodium lauryl sulfate (also called SLS or SDS) is an excellent foaming agent (in addition to being a great cleaning agent), therefore it is often used to increase the sudsiness of a shampoo. Unfortunately, sulfates bind so well to the oils in your hair that a high concentration of sulfate can actually remove too many of your natural oils and ultimately dry out the hair strands. If you have wavy/curly hair and you've been told by your hairstylist to avoid sulfates at all costs, now you know why.

Shampoo doesn't need to be this complicated though. My favorite shampoo is primarily comprised almost entirely of water, glycerin, and a few aromatic molecules. The glycerin makes the shampoo viscous (a thick liquid that is the opposite of *runny*), while the aromatic molecules make the shampoo smell good. There is some research to show that the fragrant molecules dry out your hair, but I like shampoos that smell like flowers so I've decided to take the risk. But honestly, your shampoo needs something to bind with the grease so even the best shampoos contain a few drops of sulfate (or another surfactant) to pull out the oils from your hair. If you are looking for a new shampoo, check out any of the Davines products. They don't contain any parabens or sulfates, and the company itself is committed to environmental sustainability and philanthropy.

Of course, we can all counteract the drying effects of shampoo by diligently using a conditioner (and a deep conditioner) afterward. Marketing campaigns give the impression that conditioners are intended to soften or moisturize our hair. And while this is not entirely inaccurate, these claims just happen to be side effects of these conditioners' intended purpose: to minimize friction between hair strands, especially for combing or brushing out.

But how does this work? Most conditioners contain positively charged surfactants called cationic surfactants. These lab-made surfactants are large molecules that contain quaternary ammonium compounds (or quats for short). These have kind of a diamond shape, with a nitrogen atom in the center of the molecule that is bonded to four different hydrocarbons. They are the perfect example of the tetrahedral molecules we discussed in Part I.

Due to nitrogen's position on the periodic table, we know that this element is happiest when bound to three atoms. However, in certain circumstances, nitrogen can actually bind to *four* different atoms, like in a quat, giving the molecule an overall positive charge (and making it a cation). Hence, the name cationic quats.

When the quats bind to the surface of the hair, they form a powerful hydrophobic coating on the exterior of the hair. The term hydrophobic means *afraid of water*, and this coating gives us three fantastic results. The first is that the hair is easier to comb because we have reduced the friction between the strands. They are now coated in smooth quats that allow the hairs to slide right past each other instead of forming hydrogen bonds with the water. Second, the hair is softer and thicker because we have just coated it with a protective layer. And third, due to the electrostatic interaction between the hair and the coating, there is an instant reduction of "devil horns."

Devil horns are what my sister used to call flyaways before she had any understanding of static charge—a charge that the new cationic ions have now balanced and mitigated.

When quats are used properly, they can wholly recondition (get it?) damaged hair. Here's how it works: when your hair becomes mangled due to environmental encounters or extreme chemical treatment, there tends to be a buildup of negative charge at your hair ends. The positively charged quats in the conditioner are then attracted to the most damaged sections of your hair, forming strong electrostatic attractions with the negatively charged hair tips. This beautiful ionic interaction ultimately repairs the damaged ends and reduces their tendency to lift away from other hair strands, giving you lustrous, smooth hair.

The best part about all of this is that cationic surfactants are plentiful, easy to make in a lab, and relatively inexpensive. Unfortunately, sometimes companies refer to cationic polymers as cationic surfactants, which isn't entirely accurate, and we definitely do not want to use cationic polymers on our hair!

Why? Because cationic polymers (which are much, much bigger versions of cationic quats) often have high charge densities, which implies that there is a large positive charge in a relatively small space. Molecules with large charge densities cause the aforementioned product buildup (just like silicone). The cationic polymers are *too* attracted to the hair and will cling to the hair follicles and never let go. This can cause what we call over-conditioning, ultimately leaving you with that heavy coated feeling (a.k.a. greasy hair). So most of us want to avoid cationic *polymers* at all costs.

By the way, the only difference between a standard conditioner and a deep conditioner is that the deep conditioners are stronger, thicker products that sit on the ends of our hair for an extended period of time. This gives the molecules more time to complete their chemical reactions, making our hair feel healthier and much softer afterward. They work great.

It's not snake oil, it's science!

While I'm deep conditioning (or just regular conditioning) my hair, I like to use a fragrant shower gel on the rest of my body. These bodywashing products are very similar to our shampoos, but they tend to contain a higher concentration of surfactant and perfumes. My favorite bodywash is primarily water with SLS (to form bubbles and clean my body), glycerol (to make it a thick liquid), and lots of fragrance molecules (so that I can smell like sunshine). My husband's favorite has a similar composition, but since his is significantly less expensive, it's not nearly as viscous or as fragrant.

If you use a shaving cream in the shower, you will also likely use a combination of glycerol, SLS (sulfates!), and water. However, the major difference is that shaving cream is technically a foam (gas trapped inside a liquid). The action of pushing the button on the top of the shaving cream aerosol can works like this: the gaseous molecules at the top of the can are squeezed into the water-SLS-glycerol mixture at the bottom of the can. This motion pushes a fluffy foam up through a tube and out of the can—just like when we use a straw to blow bubbles in a drink. The resulting foam can then be used when shaving.

And just like sunscreen protects our skin from the sun, shaving cream protects our skin from the razor. The foamy substance forms a protective layer between the skin and the blade, which inevitably minimizes any friction that causes painful razor burn and those unattractive red bumps. The science is simple here, people, shaving cream = smooth legs.

After the shower, the first thing I do (besides toweling off and putting on lotion) is to spray a thermal protector on my hair, which is one of the most fantastic inventions of the twenty-first century. The word *thermal* means heat, so these products are commonly referred to as heat protectants. Our hair undergoes a lot of stress during the drying and styling processes. Therefore we can protect it by coating it in a thin material—just like we use oven mitts to avoid burning our hands. The thermal protectors are usually big molecules that adhere to our hair and have extremely high tolerances to heat.

Once I've sprayed my hair in the heat protectant, I typically use my blow-dryer to dry my hair. Since wet hair is "wet" because it is covered in water molecules, the H_2O molecules can evaporate naturally at temperatures below 100°C (212°F). This is probably not too surprising, for those who air-dry their hair. From a scientific perspective, letting something air-dry is identical to allowing water in a glass to evaporate.

Interestingly, when our wet hair is exposed to heat, it can react in a few different ways, since its chemical reactivity depends on the temperature of the heat source. Physical damage to the hair surface can occur at temperatures under 110°C (230°F), which would be similar to a topical burn you may experience if you hold your hands too close to a hand dryer in a public restroom. But usually the hair can recover.

Irreversible heat damage (or chemical damage) occurs if the heat is increased to 176°C (349°F or higher). This kind of heat causes immediate decomposition of the keratin chain, which you can view by checking out #hairfail videos on YouTube. These poor kids burn their hair off with hot tools because they either (1) forgot to use a thermal protector or (2) applied too much direct heat for too much time.

Generally speaking (but not, as I'll explain, in practice), the optimal temperature for hair drying is around 135°C (or 275°F). This "ideal" temperature provides enough heat to speed up the evaporation process, but not so much that it could do any chemical damage.

When I first learned about this process, I immediately assumed that most blow-dryers would deliver anywhere between 100 and 135°C of heat (212–275°F). However, after thinking about it for a few seconds, I realized how dangerous that would be. Water *boils* at 100°C (212°F), and the vapor alone can give you a nasty burn. My husband recently burned two fingers from steam when cooking, and the blisters were so bad that I almost had to take him to the hospital.

Imagine how painful it would be to have steam at 135°C (275°F) anywhere near your face. Yikes. At least we have calluses on our hands to protect us from any accidental interactions with heat. We don't have anything like that on our scalp. In fact, the skin on our head is so sensitive that 135°C (275°F) heat would be unbearable.

Instead, most commercial blow-dryers can reach heats of 40 or 50°C (104–122°F). That's pretty weak when you compare it to a heat gun, the tool we use in lab that is similar to blow-dryers, but which reaches a temperature of 593°C (1100°F). You would not, under any circumstances, want to use a heat gun on your hair. In a panic, I once tried to dry my clothing in grad school with a heat gun after being caught in a campus rainstorm. The hot air felt so good until I realized that I was *melting* the synthetic fibers from my shirt onto my skin.

So, if blow-dryers only provide a small amount of heat, how do they remove the water from our hair? Well, to answer that question, I need to break down what temperature actually is.

When scientists use the word *temperature*, we are trying to indicate the average kinetic energy of a system. In chemistry, kinetic energy describes the movement of the molecules, and it is directly proportional to their velocities—how fast they are moving.

If we were to look at a random sample of hot water, we would notice that the molecules were moving at similar speeds, but not technically at the same speed. This is a result of a number of different factors, but primarily due to the collisions between the molecules themselves and between the molecules and the container. When the water molecules build up enough speed, they can transition into a gas.

We can imagine this situation by thinking about kids in a gymnasium. If you give them free time, most would run around in pure chaos, smashing into each other or bouncing off one another without any damage to themselves. Some kids are sprinting, while other kids are lollygagging, even standing still. In this system, it would be inaccurate to say that all kids are running or all walking. Instead, we can report the average speed of the kids, which is essentially the temperature of the gymnasium.

But what does this have to do with blow-dryers? Well, let's look specifically at the kids sprinting around the room. They are obviously moving faster than all of the other kids, and they likely want to get out of the confined space so that they can truly run free. When the opportunity presents itself (like if the gym door opens), all of the kids will head for the door. The sprinting kids will be able to escape the fastest, followed by the rest (if they increase their speeds).

The same thing happens with water molecules on wet hair. The blow-dryer provides a little extra energy for the water molecules to really start vibrating. If they build up enough energy, they can let go of the hair strand and jump into the air. When all the water molecules do this, we are left with dry hair.

However, if the blow-dryer does not reach dangerously high temperatures, why do we add a thermal protector before drying? The truth is, we don't need it to dry our hair at all! The protection is actually needed for the hot tools afterward, such as straighteners or curling irons, which run much hotter. So why do we apply the thermal protector first thing out of the shower? Because it's a lot easier to apply a liquid heat protectant *evenly* to wet hair instead of dry hair.

Unlike the blow-dryer, hot styling tools can apply a serious amount of heat directly to your hair, which might cause unwanted chemical reactions if you are not careful. Quality thermal protectors simply buy you a little time to use a higher heat setting to style your hair, just like cryogenic gloves allow me to tolerate the freezing temps of liquid nitrogen in the lab—just for a few seconds though!

With the proper technique, a hot tool can provide enough heat to trigger different arrangements of the molecules in the keratin, but not enough heat to cause any chemical changes. In other words, the quick application of heat will change the interactions *between* the molecules instead of changing the bonds *within* the molecules.

In straightened or curled hair, the hydrogen bonds (IMFs) between the cystine molecules have just changed. For example, a hydrogen atom on cystine A used to be attracted to a nitrogen atom on cystine B, but now, due to the heat energy being introduced, it is more attracted to the nitrogen atom on cystine C. This rearrangement of molecules may not seem like a big deal, but it is similar to replacing all of your red Legos with blue Legos in a Lego fort. The building blocks have the same structure and the same strength, but your fort has a different physical appearance. The new bonds formed within our hair allow for us to convert straight hair into curly hair or curly hair into straight hair. The direction of the styling tool and the heat application move the molecules into a position.

The heat is needed to form new bonds between the molecules. But the hardest part—for me anyway—is waiting for your hair to cool. This is the setting process, and it allows for the new hydrogen bonds to settle into their new positions. If you mess with your hair before it has completely cooled, you'll break these hydrogen bonds and reset your hair to its natural state. Since these bonds are temporary, a shower will reset the new bonds too.

To avoid the frustration of spending forty-five minutes on your hair just to have it ruined by some wind, many people favor hair styling products, such as hair spray or mousse. The main difference between these two is that the hair spray is often a liquid with an ethanol base, whereas the mousse is usually a foam. Other than that, they're quite similar! Both products (typically) contain a thin polymer that causes the hair strands to stick to one another. I hope this doesn't gross you out, but I always picture it like tiny spiderwebs between my strands of hair. When the hairs get close together, they get caught in each other's web, ultimately locking our curly/straight hairstyle in place.

There are different strengths of hair sprays available, and their relative strength correlates directly to the size of the primary polymer in the product. As suspected, larger polymers have a stronger hold because there are more atoms available to form thin films with the neighboring atoms. However, the larger molecules require larger droplets, which are bad for two reasons: (1) they take more time to dry, and (2) they leave your hair feeling hard and sometimes even sticky.

The cheapest products typically bind well to your hair and provide a strong hold, but they can contribute to that pesky buildup. In general, smaller polymers feel much more natural in your hair and they do not cause buildup, but they are not as strong. That's why they are usually used for hair sprays that give you a flexible hold.

Most hair sprays use an ethanol/water mixture to deliver the materials to your hair. The water is used to keep the polymer in solution. Otherwise, we would have to apply a thick rubberlike material to our hair (gross!). Ethanol is used because it has a relatively high vapor pressure, which increases the hair spray's rate of evaporation once it's applied.

Remember that water would ruin the new hydrogen bonding, so scientists had to come up with a different way to add the polymer to our hair. Each styling product has its own unique polymer:water:ethanol ratio that gives it its specific strength. I have extremely fine hair so I typically use a flexible hold hair spray, pre-curling iron, to encourage the hydrogen bonds to form, and then a strong hold spray post-curling to lock them into place.

Once I'm showered, blow-dried, and done styling my hair, I move on to my favorite part of getting ready: makeup. I start with a primer, which is a step that truly should not be skipped. A good primer acts like two-sided tape: it forms IMFs with your skin *and* your makeup, and helps to lock your makeup into place. The reason we need to wait one to two minutes to let the primer dry is so that we do not accidentally wipe any primer molecules off during the application of the next layer—foundation.

Depending on your skin type, there are a number of different types of foundations that you can use: liquid, powder, tinted, and whipped (to name a few). I found a tinted moisturizer that I like to use because it gives me an even skin tone, but it also hydrates my skin. Dry skin can be uncomfortable; it's itchy, and it can cause scaling or cracking. Perhaps you won the genetic lottery, with skin that holds more water, but for me, moisturizer is essential!

So, what is moisturizer anyway? By definition, it can be any product that claims to enhance skin appearance. In actuality, it does not have to truly fix any skin conditions, but the best ones either improve dry skin or combat photoaging (or both).

Most dermatologists agree that dry skin is a result of four major issues. First, there is a lack of water in the outermost layer of skin, called the stratum corneum. The purpose of the stratum corneum is to provide the first layer of defense from a bacterial invasion, which is done best when the skin cells are properly hydrated. If the cells are full of water, they look "swollen" and reflect sunlight evenly—giving you perfect model-like skin. But when the cells lose water, they shrivel up and give you an unflattering skin texture.

Another factor that contributes to dry skin is when there is a high turnover rate for the epidermis. This happens whenever the skin cells easily and quickly replace themselves. If this process occurs too quickly, the skin cells can have a disproportionate water content. There is also evidence of dry skin being attributed to a disrupted lipid synthesis. The ideal epidermal surface lipids contain 65% triglycerides, diglycerides, and free fatty acids and 35% cholesterol (by weight). Any deviation from this ratio will negatively affect the production of sebum (grease). This is just a really fancy way to say that fewer oils = dry face.

The last variable that contributes to dry skin is the most obvious. Any time you damage your skin, for instance you've cut or scraped yourself, you break cells causing skin barrier damage. This can lead to an infection if you are not careful, but it can also give you dry, itchy skin during the cell restoration process. Any of these issues might cause a person to reach for the moisturizer.

As a former Michigander that always fought dry skin in the winter, I swear by unscented Jergens Ultra Healing lotion. It contains petrolatum—you probably refer to it as petroleum jelly—which is simply a gel formed by several different hydrocarbons. These nonpolar molecules absorb into the top layers of your skin and repel any polar molecules that come near them. This means that any polar water molecules that try to escape from your body will be repelled back into your skin, which ultimately make your skin feel nice and hydrated.

But which moisturizer is right for your skin? It's honestly a matter of personal preference. For example, I can't stand the foundations that leave my face feeling greasy, yet some of my friends will swear by those creams. The truth is, as long as you use a quality product regularly, you will begin to notice a difference in your skin's texture. Any proper moisturizer will essentially form a protective layer over your skin, preventing cell water loss.

Optimal skin hydration will also combat photoaging (or the early aging of the skin caused by sun exposure). When cells are fully hydrated, we can perceive the difference in our skin's firmness and elasticity by the smoothing of lines and wrinkles. More importantly, we can actually quantify these changes by testing the skin for surface conductivity or surface extensibility.

Did you ever have a biology teacher ask you to examine skin cells using scotch tape? Surprisingly, you can learn a lot about your epidermal turnover rate just by examining the cells removed from your skin with a piece of tape. If the tape can be removed easily, then you have what is referred to as a normal skin type. If the tape slides around at all, then you have oily skin. And if you're getting a LOT of cell removal (trust me, it'll be obvious), you might want to think about changing moisturizers. When I do the tape test on my body, my legs are classified as dry and my forehead is oily (which is why I do not use petrolatum on my face).

Now on to the fun stuff: blush, bronzer, and eyeshadow all work similarly to foundation, where the makeup forms IMFs with your primer (and sometimes your foundation) to adhere to your face. The different colors in our blushes and bronzers come from neat molecules like carmine for reds, tartrazine for yellows, and iron oxides for browns.

Ancient Egyptians also used carmine—from smashed bugs—to give their lips a beautiful red hue. Thankfully, nowadays our lipsticks are made with dyes instead of bugs, in addition to molecules like bismuth oxychloride (to give a white, frosty appearance) or titanium dioxide (to lighten red dyes into shades of pinks).

What most people don't realize is that there is a lot of science packed into the application process of lipstick. Let's start with the classic shape of lipstick. When you open a tube of lipstick, you will see a slanted edge on top of a sturdy, cylindrical tube of wax. Most manufacturers use a form of carnauba wax (or palm wax) to give the lipstick some strength to hold its shape, otherwise when you push the lipstick against your lips, it would smash into a flat pancake.

In addition, they also spike the mixture with petrolatum (or olive oil) to help the dye transfer from the lipstick to your lips. This is one of the main reasons why lipsticks are usually soft and smooth. This texture is also derived from the silicone oil that is used to lock the dye into place on the lips. When partnered with petrolatum, this dynamic duo can give us the desired ultra-long-lasting lipstick that will last all day.

Interestingly, mascara is basically a variation of lipstick, but for your eyelashes. Like lipstick, it was developed by the Egyptians, contains titanium oxides to lighten the color (for the less-intense shades), and uses carnauba wax to give your eyelashes structure. In order to make mascaras waterproof, they add a big, nonpolar molecule called dodecane that repels polar water molecules. Without this hydrocarbon, water will dissolve your mascara and make it run down your face.

The biggest difference between mascara and lipstick is that mascara may also contain nylon or rayon. Nylon is a very large polymer, and it is used in lengthening mascaras. There are two classifications of polymers: natural and synthetic. Natural polymers are polymers that we can find out in nature, like cotton (and they are even in the DNA in our bodies). Synthetic polymers, like nylon, rayon, and polyester, are polymers that are made in a lab.

Synthetic polymers—sometimes called plastics—are made up of smaller repeating units of varying patterns, kind of like a chain of paper clips. Each paper clip is unique and separate, but they are joined together through a minor connection at each end. The molecules are held together through covalent bonds—both *within* each molecule and *between* the repeating molecules.

This paper clip chain of molecules forms big (but thin) fibers, which are stacked upon each other. These stacks have their own strong dispersion forces between molecules, and when the stacks get big and numerous enough, they create collections of polymers. When the right molecules are put together in the right order, like in the case of nylon, the resulting polymer can be quite strong and stretchy. However, if you've ever gouged a hole in your pantyhose (which is also made of nylon), then you know that polymers can be pretty weak too. The strength of the polymer depends on the bonds of the molecules, and the IMFs keeping those bonds intact.

Polyamide—the polymer used to make nylon—has repeating molecules linked together in a unique way called an amide connection. These amide connections are very specific: they happen when a carbon atom on one side of molecule A forms a covalent bond with a nitrogen atom on the opposite side of molecule B. The catch is that molecule A and molecule B are actually identical, so this connection repeats itself and creates a line of molecules that are connected through these super strong carbon-nitrogen covalent bonds.

POLYAMIDE BY ANOTHER NAME

Have you ever heard of a scientist named Stephanie Kwolek? She's an American chemist that passed away in 2014 after spending more than forty years working as an organic chemist at DuPont. Back in 1964, she was working on a new molecule that could replace steel in racing tires when she accidentally formed a bizarre solution in the lab.

Kwolek was so intrigued by the half-liquid, half-solid substance that she asked a colleague to run the material through a spinneret, which is a piece of equipment that tries to spin solutions into fibers. If the experiment was successful, long needlelike fibers would be produced that look just like glass wool. Luckily for Kwolek, that's exactly what happened for her material. She was so happy with this result that she decided to test the new molecule for robustness, and to her surprise, found out that it was five times *stronger* (by weight) than steel.

After a few more experiments, Kwolek and her colleagues learned that the new substance could become even stronger after heating it. For those of you who have never worked in a lab, this result would be like watching Superman walk into a fire before miraculously turning into the Hulk. Somehow, the heat from the fire was forcing the molecules to rearrange in such a way that it gave the substance superhero strength.

The material Kwolek discovered is Kevlar. Today, we use this molecule in all sorts of things, from bulletproof vests and fiber-optic cables to the spacesuits astronauts will wear on Mars. This giant molecule is called poly-paraphenylene terephthalamide, and it is a synthetic fiber.

Kevlar is one of the strongest materials known to mankind. The atoms are packed so tightly together and bonded so strongly to all of its neighbors, that nothing—not even a bullet—can break them apart. In the Orlando nightclub shooting, one of the police officers that responded to the call was saved by his Kevlar helmet that stopped a bullet from entering his head. And in the Parkland shooting, a group of high school students hid behind Kevlar sheets they found in a Junior ROTC room.

This fabric saves lives. All because of the (exceptionally strong) attractions between molecules.

The elastic fibers in the polyamide were first invented in the 1930s, and were immediately recognized as excellent candidates for clothing (not makeup). For example, when nylon stockings were first released in 1939, they were a huge improvement over stockings made from cotton or wool. Women would wait in long lines to purchase just one pair of stockings—kind of like how lines form for Black Friday sales now.

Just like other fabrics, nylon is pulled into long, thin fibers before being grouped together like planks in a fence. Then those fibers are knitted together in an intricate pattern of loops to produce nylon fabric. While the fabric is very stretchy, it does not breathe as much as the polyester fabrics we discussed in the previous chapter because the molecules are so tightly bound together.

As mentioned, when these new nylon stockings were first produced, women of the time were obsessed with the groundbreaking fabric. During World War II, DuPont shifted their production and instead of using the polyamide material to make stockings, they started to make parachutes for the American military. Because of this, the supply of the stockings went down, and the demand for them went up, causing—and I am not joking here—the nylon riots. Women were so upset that they could not get their hands on the nylons that they started to fight each other for hosiery. Some women were even robbing their neighbors!

After the war, manufacturing companies began producing stockings again, this time by blending the nylon fabrics with other natural and synthetic fibers, like cotton and polyester. These blended fabrics were a brand-new idea at the time, and they proved to be extremely popular in women's fashion. The new stockings were lightweight and elastic, in addition to being inexpensive and cute. But from a molecular perspective, they were just a different type of polymer.

Nowadays, we commonly use all types of polymers in our fabrics. You can commonly find nylon blended into your outdoor clothing, like your rain jacket or waterproof pants. One of the most inexpensive polymers is called polyethylene terephthalate (PET). It is the fourth most produced polymer in the world, and you probably already know it by its street name, polyester. And just like nylon, polyester (and other common fabrics) is made with various polymer chains and binding mechanisms.

I could go on and on, because the truth is that every item in your closet is loaded with chemistry—your velvets contain acetates, your cottons come from cellulose, and some of your wicking clothing uses a polymer called polylactide.

Even your jewelry is chemistry! Your earrings, bracelets, and necklaces are just metals on top of other metals, and then melted to form new shapes and textures. Speaking of which, grab some big earrings and your favorite teeny, tiny bikini because it's time to head to the beach!

8

GIVE ME SUNSHINE

At the Beach

In Austin, we can get to the ocean in just under four hours. All we have to do is throw the dogs in the back of the car, open the sunroof, and enjoy a sunny drive to Galveston or Corpus Christi. In the summer of 2019, being able to get to the water was almost a means of survival. In August of that year, we had temperatures of 100°F or higher for twenty days of the month. It was brutal.

My husband and I practically lived at the beach that summer, which was when I started to notice the chemistry in action by the ocean. From the avobenzone in my sunscreen to the polymers in my swimsuits, I could spot a real-world example of my favorite science in every direction.

Case in point: my cooler.

The one thing we absolutely relied on that summer was a quality ice chest to keep our food and drinks insulated and miraculously cold despite the sizzling heat. Our favorite one is made from polyethylene, but polystyrene can be used too.

I'm guessing you haven't thought much about the science of coolers, but it's pretty astounding; these workhorses use a special molecular structure to literally trap cold air inside. There are two different forms of polyethylene commonly found in most big, sturdy coolers so let's look at that polymer first. Just like the name implies, polyethylene polymers are comprised of many ethylene molecules, and it is currently our most common form of plastic. Ethylene is a hydrocarbon with the formula $H_2C{=}CH_2$, and it is an extremely flammable gas. It's a nonpolar species (the electrons are evenly distributed across the molecule), which means it can only form dispersion forces with its neighboring molecules.

However, under extremely high pressure, ethylene can react with itself chemically to form a huge ethylene chain. When this happens, the double bond breaks, leaving a single bond that holds the original molecule together. After this step, the carbon atoms can form new covalent bonds with different carbon atoms, resulting in a long hydrocarbon chain.

Let me break that down using the Ryan Reynolds example from the first part of this book. If you remember, we discussed how I could form a double bond with Ryan by holding both of his hands. But if I want to go through polymerization reaction, I need to release one of Ryan's hands and reach out to form a new connection with another adorable celebrity, like Joe Manganiello. Ryan, of course, will do the same thing, therefore he can produce a new bond with Blake Lively.

This domino effect will continue until all carbon atoms are surrounded with four covalent bonds (because of its position on the periodic table). The product, polyethylene, is nonpolar, with dispersion forces between the polymer fibers. These interactions are very similar to the IMFs formed between individual ethylene molecules.

These molecules are gargantuan with molecular weights of 10,000–100,000 g/mol. Since polyethylene is a huge, nonpolar molecule, it is not soluble in water, which makes this polymer a perfect molecule to use in the production of an ice chest. It's insolubility in water allows us to fill the chest with ice and bring it to the ocean.

Polyethylene is also used in sandwich bags, which protects our sandwich from getting wet and soggy from the ice! But what is the difference between the polymer used to make an ice chest and the one in our sandwich baggies?

For starters, the exterior plastic of most coolers is made from high-density polyethylene (HDPE), while the sandwich bags are made from low-density polyethylene (LDPE). Let's talk about LDPE first, so I can explain the difference. LDPE was first used back in the 1930s, and it has a lower density than HDPE (obvious, I know). But even though the two plastics have the exact same atoms and covalent bonds within the polymers, the *way* in which the bonds have formed is very different.

The LDPE is a polymer made from covalent bonds between neighboring ethylene molecules. Like we discussed before, when the ethylene molecules start reacting with each other, the double bond breaks and a new single bond forms between nearby carbon atoms. This process forms what is referred to as a branched hydrocarbon chain. So, instead of forming one orderly, single-file line of carbon atoms, the bonds are made between random carbon atoms to form T's all over the structure. A linear shape would be nice and orderly, like a chain of kindergarteners heading to lunch, while the branched shape is a chaotic mess, like kindergarteners at recess.

The branched shape of LDPE makes it much weaker than HDPE polymers (and a lot more stretchy). Because this shape makes it difficult for the molecules to pack closely together with the other LDPE polymers, they cannot form strong dispersion forces. This means that there is a lower-density of polymers in a given space, specifically, a density range of 0.917–0.930 g/cm^3.

What that *really* means is actually quite significant, and it's why we use these types of polymers so often in our daily lives. They are stretchy and resilient, which makes them perfect for sandwich bags (and plastic beach bags—before everybody realized they created a lot of excess waste, and converted to reusable bags). LDPE polymers are ideal for packing big sandwiches because they can be wrapped around the fragile bread, yet still keep your lunch safe from moisture, especially when compared to something natural, like paper, that's completely rigid and does not stretch or bend or shield from water.

Technically, polyethylene is not considered a hard or rigid polymer, which means that, in theory, we should be able to mold the molecules into different arrangements. Our ability to change the shape of these plastics has to do with how the molecules are lined up within the polymer. For example, if we grab two sides of a plastic sandwich bag and pull, we will see the entire shape of the bag change. It will immediately adapt to the stress by increasing in length and decreasing its width, basically forming the shape of a dumbbell. If we pull too hard, the plastic will eventually snap.

This process, called *necking*, occurs whenever the molecules within a polymer try to adjust to stress. Before we started messing with the plastic bag, the molecules were arranged in a haphazard manner, like wet pasta in a pan. But as soon as we start to pull on the plastic bag, the molecules began to straighten out. Like if we were to zap our pan and change all of the wet spaghetti into dry spaghetti in an instant. With stress, the molecules change shape from bent to straight, and all line up in perfect order. The combination of the long, thin shape and the organized arrangement of molecules allows for the plastic to be stretched out, forming the middle part of the dumbbell shape I mentioned earlier.

But if you release the bag, it'll mostly go back to its original, chaotic shape. (There may be some damage to the spots where the plastic connected with your fingers, but otherwise, the polymers on the inside should return to their initial arrangement.)

Unlike LDPE, HDPE has a linear configuration with strong covalent bonds between hundreds of carbon atoms. I'll explain how this is done below (it's pretty wild). But what you need to know is that this shape makes HDPE polymers much stronger than the branched LDPE polymers because HDPE polymers can closely pack together, like uncooked spaghetti noodles. With this configuration, the polymers can form strong dispersion bonds between each polyethylene molecule, resulting in the high-density polymer used in big ice chests with a density range of 0.930–0.970 g/cm^3.

While scientists were able to predict that HDPE would be stronger than LDPE, they initially struggled to figure out a quality way to synthesize the polymer. Twenty years after the discovery of LDPE, a German chemist named Karl Ziegler started messing around with ethylene. After every reaction, he found a version of the same product, butene, a molecule with all single bonds except for one double bond.

Ziegler was captivated by this unexpected chemical reaction and immediately started to conduct more complicated experiments. He learned that there were tiny amounts of nickel hiding somewhere within the ethylene gas that created the formation of butene.

Ziegler freaked out and began throwing all kinds of metals into his ethylene mixture (one at a time, of course). It was all a little haphazard, but he got results. He immediately noticed that zirconium and chromium would also produce a mixture of polymers, but that titanium worked the best to produce polyethylene in the desired linear shape.

This was a groundbreaking discovery because metals had not yet been used to form a covalent bond between two other molecules. Before this, scientists were just throwing molecules together and changing their concentrations, pressures, and temperatures to try to encourage the right chemical reactions to occur. But without knowing it, Ziegler had just triggered a new type of catalysis—a huge topic in my field, inorganic chemistry.

When Ziegler presented his work at a conference in 1952, an Italian chemist named Giulio Natta was convinced that he could help take these "Ziegler catalysts" to the next level by adding something else—a cocatalyst. And that's exactly what it sounds like: a second metal added in, to help the first provoke a chemical reaction.

It turns out that Natta was right, and they quickly developed the Ziegler-Natta catalyst, which has become the generic name for any two cocatalysts that convert double bonds into single bonds, specifically by producing a long polymeric chain. This unorthodox synthetic method was so innovative that it kick-started an intellectual frenzy over new polymer research. Because this discovery revolutionized how polymers were made at the time, Ziegler and Natta won the Nobel Prize in Chemistry in 1963.

With a quick and easy way to generate HDPE, engineers were eager to begin using the strong polymers for common household materials. Nowadays, HDPE is not only used for coolers or ice chests, but also in boats, beach chairs, water jugs, and the container for your sunscreen. It is literally everywhere at the beach, just like LDPE, which is found in the playground slides, your Tupperware lids, and juice box containers.

Over the years, HDPE has become a favored molecule for plastic coolers because of its impressive insulating properties. As an insulator, the polymer minimizes the amount of heat transferred across a substance. When coupled with a hard foam, the sunlight has a difficult time penetrating the exterior of the ice chest. As a result, the hot air stays on the beach, and the cool air chills your six-pack in your cooler.

Did you know that in the 1960s, manufacturers started using polymers for six-pack rings as a substitution for metal and paper holders? I imagine the change was well received since the paper holders immediately disintegrated from the condensation on the cans. However, in the late 1970s and early 1980s, there was a huge environmental campaign to ban any plastic six-pack rings that were too rigid. At the time, small wildlife animals were getting stuck in the rings and were unable to get food, so manufacturers quickly switched to more flexible plastics as a way to avoid any unnecessary animal fatalities. This meant that the new six-pack rings needed to be strong enough to hold the cans together, but weak enough to be ripped apart. LDPE was the perfect candidate for this role.

In 1993, the Environmental Protection Agency (EPA) demanded that all plastic rings must be biodegradable; the plastic must be able to break down naturally without any human assisting with the process. The solution was to use photodegradation, which is a process where UV light breaks down the bonds in the polymer (more on UV light in a bit). Depending on the size of the plastic, this entire process can take months, if not years. However, if the plastic is thrown into a landfill and buried beneath other trash, then the sunlight will not be able to activate the mechanism, and the plastic will never be able to decompose. Polymers with high densities, like HDPE, have an even harder time decomposing by themselves due to their large sizes.

This is true for all ethylene-based polymers, including polystyrene (PS), the large molecule used to create foam coolers.

Polystyrene is made from the molecule styrene, which looks a lot like ethylene. The only difference is that the terminal hydrogen atom on ethylene ($H_2C{=}CH\mathbf{H}$) has been replaced with a big, bulky, benzene ring (C_6H_5) to look like this: ($H_2C{=}CH\mathbf{C_6H_5}$). And just like how polyethylene had hydrogens all over its structure, polystyrene has a six-membered benzene ring (C_6H_5) projecting from every other carbon atom. It's a *big* molecule.

Polystyrene was first synthesized in 1839 by a German apothecary Eduard Simon. He extracted the resin from an oriental sweetgum tree and boiled it down to an oily substance. Through a purification process, he was able to isolate a molecule called styrol. This molecule eventually thickened into a goopy substance, like jelly. Later, in 1866, a French chemist named Marcellin Berthelot identified the molecule as a long chain of hydrocarbons with alternating benzene rings. But at that time, scientists had yet to discover polymers. Therefore, it took another eighty years for this substance to officially earn its name, polystyrene.

The three common structures that polystyrene can form are called isotactic, syndiotactic, or atactic. This term indicates the *tacticity* of the polymer, which is a fancy way to indicate the position of the benzene rings. If all of them are on one side of the polymer, then we use the term isotactic. From a chemistry perspective, it doesn't matter if all of the benzenes are on the right or the left side. We just care that they are all on the same side—like if a centipede only had right legs. These polymers are the strongest of the three because they can nestle in closely with all the neighboring polymers.

For this reason, the isotactic configuration gives us the best foam cooler. Just like with the polyethylene coolers, the hot air is unable to push through the foam to get to the cold sodas and fruit salad. However, there are two other conformations that can be found in less expensive foam coolers.

If the PS polymer has benzene groups that alternate (right-left-right-left), then we use the term syndiotactic. These polymers look like the stem of a rose, where the leaves alternate sides all the way down the stem. With this shape, the polymers can no longer line up like uncooked spaghetti noodles. The leafy part of the molecule, quite frankly, gets in the way.

If the benzene groups seem to be randomly placed or unorganized, we use the term atactic. These polymers are the weakest of the three, therefore they have the lowest melting point. For this reason, the atactic polymers are the most flexible of the three structures, and that gives them a more rubberlike texture.

Because the benzene rings are so big, PS polymers are usually in the form of an atactic structure, meaning that the benzene rings are randomly placed across the polymer. Some are on the left, and some are on the right. Some are next to each other, while some are evenly spaced out. There is no pattern.

For this reason, we really only interact with two main forms of polystyrene. The first form, called crystalline, is commonly found in single-use plastics of any tacticity, like in the plastic forks and knives you may pack for your picnic on the beach. And if you prefer to use food wrap—commonly called by the brand name Saran wrap—instead of sandwich bags, then your cooler will be filled to the gills with polystyrene.

Food wrap is one of the coolest products ever invented because it can seal a container shut without using any type of adhesive, like tape. The intermolecular forces are so strong in polystyrene that that atoms are attracted to each other, which encourages the plastic to stick together. If you ever have a piece roll up on you, just slowly smooth it back over the container. You are giving the atoms a little more time to form dispersion forces between the different polymer chains.

The second form of polystyrene is expandable (hence the name *expandable polystyrene*). You commonly see this version of the polymer at the beach in anything made with Styrofoam, like foam cups or foam ice chests. (Styrofoam is the brand name of foam, just like Kleenex is often used in place of the word *tissue*.) Some government agencies even use PS underneath its roads to provide insulation so that the asphalt doesn't freeze and buckle.

Expandable polystyrene is a fluffy polymer. When industrial scientists manufacture it in big warehouses, they go through a really fascinating process. First, they divide the polystyrene molecules into small round pellets like caviar. The pellets are then pumped with air before being dropped into a large mold. From there, the pellets are steam-fused together, which means that they are heated to a high temperature before being squeezed together into the shape of the mold.

PS foams are around 3–5% polystyrene—and the rest of the product is just air. That's why most Styrofoam products, like foam coolers, are so light. PS foams are also great insulators, which is why we can use the polymer to keep our food and drinks nice and cool.

But neither crystalline nor expandable polystyrene is strong enough to withstand the pressures of a carbonated beverage. For this job, we have to look to our friend, polyethylene terephthalate (PET), that we discussed in the last chapter in the context of polyester.

Interestingly, this same polymer is commonly used in the production of water and soda bottles. The plastic is translucent, which is nice so that we can see what we're drinking. But the bottle is also sturdy enough to withstand the pressure built up from the collisions with all of the carbon dioxide molecules inside the soda. If you are like me and you enjoy bringing fruit to the beach, you might also have PET molecules surrounding your delicious berries too, in the form of plastic clamshell containers.

Once all of the snacks and beverages have been packed up

in the cooler, we throw on our swimsuits and jump into the car. Most swimsuit fabrics, of course, are also made of polyester polymers (or nylon polymers) with about 10–20% spandex. This allows the resulting swimsuit to be stretchy and elastic and super-duper comfy.

This should not be too surprising based on what we've already discussed about polyester and nylon, but there is something particularly fascinating about the synthetic fiber, spandex (a.k.a. Lycra or elastane). This polymer was made from a combination of polyether and polyurea back in 1952, and was originally synthesized as a replacement for rubber in women's girdles. Because of how comfortable the polymer was in undergarments, it quickly earned its name *spandex* as an anagram for the word *expands*.

All three of these synthetic fibers are ideal for swimwear because of their nonpolar properties. When we jump into the ocean, the water molecules will immediately wash into the gaps between the fibers in our swimsuit, which would saturate a natural fabric like cotton or wool with water. However, since the materials we wear are made of nonpolar materials, the polar water molecules are actually repelled by the nonpolar fabrics. While this does not completely stop the absorption of water molecules—our bathing suits are not waterproof, after all—it does minimize the amount of water that can be absorbed.

To prove my point, let's imagine a swimsuit made of cotton, which is comprised mostly of another polymer called cellulose. Cellulose is made up of a long chain of glucose molecules that are covered in alcohol functional groups (OH). This makes cellulose molecules extremely polar, and as a result it forms lots of hydrogen bonds with the water in the ocean. These IMFs are a good thing when goats are trying to dissolve cellulose in their stomachs, but on a bathing suit, these bonds are what will give you the embarrassing saggy bottom. So much water is absorbed and bonded to the cotton fibers that the molecules weigh the fabric down and pull the swimsuit off your body.

To avoid such a humiliating situation, I tend to buy swimsuits that contain some combination of nylon and Lycra fabric. And if I can purchase something that is made with recycled nylon, I will. The process of recycling clothing, called melt extrusion, breaks down old polyamide polymers with high temperatures and pressures. However, when it comes to actually purchasing swimwear, I'm mostly concerned with the fit and how well the suit will stay on my body.

My husband, on the other hand, could not care less about the fit of his board shorts. He just wants something that is lightweight and that will repel the water so it's usually a mixture of polyester/spandex blends for him. The water literally rolls right off his suits; however, he needs a drawstring to keep them in place because the fabrics do not have the same elasticity.

When we arrive at the ocean, I usually see spandex everywhere. It's in the swimwear and wetsuits, biking jerseys and beach volleyball attire, and even in many of the swimsuit cover-ups. And if you look closely, you can also find it woven into the fibers of nicer hats to give them their structured fit.

But why do we even wear a cover-up or a hat at the beach in the first place? What is it that we are trying to protect ourselves from?

Light.

Super harmful, cancer-causing, beach-saturating light.

But what is light? And is it chemistry?

Absolutely! In fact, everything you can see right now, at all times, is interacting with light. The red book in the corner is emitting visible light in the red region of the light spectrum, while your purple shirt is emitting light in the purple region. The light from your lamp and your cell phone battery are both in the form of infrared (IR) radiation, or heat (that's why they're warm). And if you have a black light in your room or your curtains are open, then you are exposing yourself to ultraviolet (UV) radiation. So, unless you are in total darkness right now, you are interacting with light.

Scientists have been investigating light for a very long time. Back when they believed that all matter was comprised of earth, air, water, and fire, a Greek philosopher named Empedocles was convinced that the element fire shot out of our eyeballs, illuminating our environment and allowing us to see.

There were obviously big flaws with this theory, chief among them the fact that with fireball-shooting eyeballs, we should be able to see in the dark. Incidentally, Empedocles was also the person who originated the idea of the four elements theory. He was wrong about both ideas; as we all know, humans do not have death-eyes like Cyclops in *X-Men*.

It wasn't until the 1600s that a French philosopher named René Descartes proposed that light behaved like a wave. At the time, Leonardo da Vinci had already discovered that sound travels in waves, therefore it was very reasonable for Descartes to postulate that light might do something similar. This idea alone completely shifted how we eventually understood *all* atomic elements—including protons, neutrons, and especially electrons—and their ability to exist in both particle and wave form, simultaneously.

When I talk about waves in this section, I want you to think about a wave in the ocean. The wave always originates from something that is giving off energy (like a boat or a Jet Ski), and then it propagates through the water unimpeded until it slams into land or bends around an island. When we talk about waves in the context of sound, that means that sound waves can bend around obstacles (like a wall) so that someone can hear the oven timer in the kitchen even though they are in the next room over. The person does not have to see the timer, or be in the direct path of the timer, in order to hear the noise.

The theory of light acting like waves held some weight at the time because it explained why light could travel through liquids at different speeds. However, if light were to behave *exactly* like sound waves, then we would expect for light to be able to bend around obstacles as well. But the problem with that theory is that we cannot see the beam of light from a flashlight through a brick wall. Even though some aspects of light seemed to behave like a wave, scientists knew that the wave theory was not the perfect explanation for light.

Just a few years later, an English physicist named Isaac Newton decided to publish a little known French philosopher's work posthumously, in an attempt to explain the flaws in the wave theory of light and discredit it. This man, Pierre Gassendi, argued that light in fact behaved more like a particle, which meant it acts like something with a mass. And in some ways, this was true—and was the basis for what we now call the photon—but this theory was also incomplete.

If light were a particle, one would expect for a brick wall to stop all forms of light from shining through, just as it stops anything with mass, like a baseball. In the same way that we cannot throw a baseball through a wall, we cannot expect light to move through (or around) a wall. This is mostly true, but doesn't explain light refraction—which is responsible for rainbows, or for light bending around the edges of a door—which shouldn't be possible if light were made up of infinitesimal particles travelling in a straight line.

To keep a very long and extremely complicated story short, I'm going to fast-forward to the 1920s, when the French physicist Louis de Broglie suggested that *all* matter behaves like both waves and particles. This theory was later adapted to include light, which was how wave-particle duality was born.

Wave-particle duality is one of the most fundamental principles in chemistry because it explains how particles (like protons, neutrons, and electrons) can operate like waves. We can use wave mechanics to predict *where* the electrons will be located in an atom or molecule, and that information tells us everything we need to know about sunlight.

Remember those orbitals that we discussed in the first part of this book? All of the *s*, *p*, *d*, and *f* orbitals are derived from one equation (the Schrödinger equation) that was based on wave-particle duality. In fact, when you solve the Schrödinger equation, the solution you get is a number (the row on the periodic table) and a letter (the orbital) for that electron. This meant that for the first time, scientists could determine the energy of the electron and its location relative to the nucleus with decent accuracy. All thanks to our friend, the Austrian-Irish physicist named Erwin Schrödinger.

This groundbreaking theory of wave-particle duality not only applied to matter, but it also seemed to be a good theory that could explain the previously mentioned properties of sunlight.

What scientists learned is that the light on Earth (and on the beach) comes from the sun in the form of electromagnetic radiation. *Electromagnetic radiation* is just an all-encompassing term that is given to any form of energy that travels (or radiates) through space in an electromagnetic field. The energy moves through space with two perpendicular waves (electric and magnetic), hence the name electromagnetic.

Since electromagnetic radiation—or electric and magnetic energy transfer—is such a fundamental principle in chemistry, I want to look at this topic a little more closely. If a certain type of light travels from left to right, the electric wave and magnetic wave have to move left to right too.

To keep it simple, we can assume that the electric field moves left to right across a tightrope. Therefore, by definition the magnetic field will also move left to right, just at a different orientation. Instead of moving horizontally across the tightrope like the electric field did, the magnetic field will travel in a vertical wave pattern alternating from above the tightrope to below the tightrope and then back above the tightrope again. When both of these things happen at the same time, light can move through our atmosphere. If a molecule (or anything really) interferes with either the electric or magnetic waves, the atoms either block or bend the light.

Quantum mechanics—the science of subatomic particles—can get extremely convoluted very quickly, so I just want you to focus on one thing: there are lots of different kinds of electromagnetic energies, a.k.a. light! Scientists have organized light into a spectrum (called the electromagnetic spectrum) that ranks them by their wavelengths. On one end of the spectrum are what we call radio waves with super long wavelengths that are about the size of a building. They are so big that they are very low in energy. Radio waves cannot hurt your body at all, which is why we can safely use radio waves for both Wi-Fi and Bluetooth.

The other end of the spectrum contains gamma radiation with super short, high energy waves. These wavelengths are closer to the size of an atomic nucleus; they are incredibly small. This kind of radiation is so dangerous that it can wreak havoc on your body and cause serious damage to your internal organs. For that same reason, highly concentrated beams of gamma radiation can be carefully used to kill cancerous cells.

But just like with Goldilocks and the three bears, there is a perfect middle ground of electromagnetic radiation, one where the waves are not too short and they are not too long, where the waves are neither too low in energy nor too high. These electromagnetic waves are in the middle of the spectrum with a medium wavelength and a medium amount of energy (relative to radio and gamma).

These waves correspond to the three types of energy that the Earth's surface receives from the sun on any given day: ultraviolet, visible, and infrared. We've already discussed visible light through dyes and fabrics and the infrared energy that's produced from the stove. Therefore, now I want to focus on the most dangerous of the three—ultraviolet light (UV)—and explain why we wear sunscreen at the beach.

UV light is the highest form of energy that the Earth's surface receives from the sun. It was first discovered back in 1801 after German-British astronomer William Herschel had correctly identified heat waves (IR). Herschel proved that energy could be emitted at a longer wavelength than visible light; therefore German physicist Johann Wilhelm Ritter started to wonder if there was another "invisible" energy but with shorter wavelengths than visible light.

Ritter started playing around with violet light (the highest energy light that we can see) and what later became known as ultraviolet light. He noticed that the "invisible" ultraviolet light was darkening a paper drenched in a silver chloride solution, much faster than violet light was able to. In fact, the ultraviolet light was interacting with the solution and changing the color almost immediately. Since violet light was the highest known energy of light at the time, Ritter knew that he had stumbled upon something of interest.

UV waves—dubbed *chemical rays* at the time—are about the size of one molecule. Even though these wavelengths are tiny, they are extremely powerful because they carry a *lot* of energy. For example, in 1878, scientists learned that UV light can be used to kill bacteria. They began to use it as a way to sterilize different products, including medical equipment, and still do to this day. In fact, in 2020, the scientific community gave a big sigh of relief when we learned that UV light could be used to combat COVID-19.

There are several different kinds of UV light (we call them UVA, UVB, and UVC). They vary slightly in wavelength, and the sun emits small amounts of all of them. Ultraviolet A (UVA) is the weakest of the three with wavelengths of 315–400 nanometers (nm). Because they have the longest wavelength within this category, this type of energy is sometimes referred to as long-wave radiation. The most common usage of UVA light in our everyday world is in the form of a black light. Even though they do not look very bright, these waves are associated with extremely high energy, which is why you should never, ever stare directly at them (or the sun for that matter). For this same reason, you should also never lie naked in the middle of a bunch of UVA rays, like in a tanning bed for instance (but more on that in a minute).

When we decrease the wavelength a little more to 280–315 nm, we move into a new category called ultraviolet B radiation (UVB). UVB light is higher in energy than UVA light, and it can be used to treat a number of different skin conditions. Psoriasis and vitiligo are two common diseases that can be helped by direct contact with UVB light. While all the symptoms may not be eliminated, they are often alleviated after exposure to the high energy light.

People that keep lizards and turtles as pets often install UVB lamps called basking lights in the cages of their reptiles. This ensures that their adorable pets are nice and cozy. Cold-blooded animals like reptiles and amphibians greatly benefit from a UVB light in their cages because their bodies can absorb the energy of their environment. It is pretty cute to watch these little creatures bask under a UVB light bulb.

But it turns out, humans need to get a little sunshine too. We all have cholesterol in our skin, which is a molecule that has four rings fused together (three six-membered rings and one five-membered ring). It reacts with the high energy from UVB radiation to produce vitamin D_3—the common name for the molecule cholecalciferol.

When humans do not get enough sunlight (and ultimately vitamin D), then they can develop a condition referred to as vitamin D deficiency that affects the body's ability to absorb calcium and lowers bone density. Ultimately, vitamin D deficiencies tend to lead to bone fractures, which is why it's important to go outside for a few minutes a day. Your skin simply needs time to let the UVB light break down its cholesterol and convert it into vitamin D.

The highest category of UV light is called ultraviolet C (UVC), and it has a range of 100–280 nm. These short wavelengths correspond to extremely high energy radiation that scientists later learned were originally responsible for the germicidal properties observed back in 1878.

And just like UVC radiation can kill bacteria and germs, they can also cause some serious damage to the cells in our body. In fact, more than 90% of skin cancers are caused by UV radiation. These "chemical rays" are so powerful that they can penetrate your skin, breaking the bonds within the molecules in what we refer to as bond dissociation (this process is exactly like what it sounds like: the bonds break and cause the atoms to dissociate).

When this happens to the molecules within your body, the newly freed atoms move around searching for new counterparts, which unfortunately, can cause new problematic bonds to form. If these bonds are produced in the wrong part of the molecule or the wrong place in the body, cancerous cells can form.

The good news, is that we can prevent skin cancer just by putting on a thick form of lotion called sunscreen. This chemical concoction absorbs both UVA and UVB radiation. But what about the most dangerous of the three, UVC?

To answer that question, I need to tell you a little bit about how our atmosphere works, and the important elements that are contained within it. The oxygen that we breathe sits in a layer of air called the troposphere, which is the first layer of the Earth's atmosphere. It contains primarily nitrogen, followed by oxygen, argon, carbon dioxide, and water—all of the gases I mentioned in the first part of this book.

The stratosphere layer is a bit higher than the troposphere and hovers just above the clouds. We actually fly in the stratosphere layer to avoid turbulence caused by the molecules in the lower troposphere layer. There are fewer molecules at that height above the Earth, therefore planes do not have to worry (as much) about varying air pressures that cause turbulence.

However, there is an extremely important layer of our atmosphere called the ozone layer that sits in the lower portion of the stratosphere. As you may be aware, the ozone layer is a very thin shield that basically acts like the Earth's sunglasses, all thanks to two molecules: oxygen (O_2) and ozone (O_3). We might not see it, but all these gases, photons, and energies are constantly interacting with each other up in the stratosphere!

When the UV sunlight hits the ozone layer, a couple different things can happen, which all depend on the energy of the incident radiation. For example, when high energy UVC light hits the ozone layer, it can break the double bond in the oxygen molecules (O=O) *if* it has a wavelength shorter than 242 nm. Anything higher than that will not be able to break the double bond. If the incident UV radiation has a wavelength shorter than 320 nm, it will be able to break the covalent bond within the ozone molecules, but not within oxygen.

So, what does this mean for the people lying out on the beach? The thick layers of oxygen and ozone work together to protect the Earth from harmful UVB and UVC radiation.

However, while they are sacrificing their molecular bonds to prevent high energy light from penetrating the Earth's atmosphere, the lower energy UVA radiation sneaks right through.

But how does this happen? Technically, UVA is weaker. Shouldn't ozone and oxygen have an easier time defending us from the dangerous UV energy? Unfortunately, no.

The only way these molecules can protect us is by letting the UV light break their covalent bonds. But the problem is, oxygen requires a wavelength shorter than 242 nm, and ozone needs a wavelength shorter than 320 nm. Anything higher than 320 nm is too weak to break the bonds in our sacrificial molecules. Therefore, the UVA light (with a range of 315–400 nm) flies right by the molecules and down toward us beachgoers.

Even though UVA light is the weakest of our three categories (with the biggest wavelength), it is the one that causes the most damage to human beings. That being said, if we were actually exposed to UVB and UVC light, it would be much worse for us, penetrating our bodies, breaking bonds, and creating molecular mayhem. We just have two superhero molecules to prevent that from happening.

If you want protection from UVA radiation, all you have to do is wear sunscreen every single day. No problem, right?

Since that's easier said than done, I highly recommend buying makeup products that already contain sunscreen. This way, your skin is automatically protected if you run outside for a few minutes or get caught chatting with your neighbor. However, if you are heading out for a day at the beach, you need to smother yourself head to toe with some form of broad-spectrum sunscreen (meaning it protects from several "spectrums" of UV light).

There are two forms of sunscreen that are quite common in the United States. One is a physical blocker, which is a sunblock that actually sits on the top of your skin. The most common version of this is zinc oxide, a thick, white cream responsible for the white-nose look of every 1980s lifeguard. Of course, my dad was the only parent in my town that made us wear the stuff—I was *so* embarrassed by it.

The second version of sunscreen is more common, and it works through a chemical process, where molecules within the sunscreen act like oxygen and ozone by absorbing UV radiation. Some of these molecules, like avobenzone, have limits on their absorption ranges, which is why sunscreens typically contain more than one active ingredient. Avobenzone is best for UVA radiation, but not so great for UVB. Octyl methoxycinnamate, on the other hand, can absorb any UVB light that sneaked by the ozone molecules in the ozone layer.

Personally, since I live in a very warm place with temperatures over 100°F in the summer, I always have sunscreen with a sun protection factor (SPF) of 30 in my linen closet. This SPF rating means that the sunscreen will block all of the harmful UV radiation that interacts with my skin, except for 1/30 of it. SPF 10 will block all but 1/10 of the radiation, and SPF 50 will block all but 1/50 of the radiation.

However, the SPF rating is only accurate if you remember to reapply the sunscreen every two hours. Otherwise, at hours three and four, your skin will be completely exposed due to a number of different factors. Not only can the sunscreen rub off in the water, but eventually all of the photosensitive molecules will decompose after absorbing the powerful UV energy.

There is some argument that sunscreen/sunblock cannot truly have an SPF over 50. Most chemists believe that we cannot have that kind of accuracy with so many variables, especially when sunscreen's effectiveness depends heavily on how diligently the person applies the thick cream and how frequently they reapply it. Personally, I have not seen enough evidence to support the claim that sunscreens can truly protect us from that much UV light, therefore I stick to sunscreens with an SPF of 30.

Just like with all man-made, chemical inventions, there can be negative side effects of certain sunscreens. For instance, even though octyl methoxycinnamate is a fantastic molecule that can protect our skin from harmful UVB radiation, it is very harmful to coral reefs. Unfortunately, most people put their sunscreen on right before they go to the beach and go straight into the water. For this reason, certain areas (like Hawaii) have placed bans on sunscreens containing any octyl methoxycinnamate (the ban will go into effect in 2021).

But it doesn't matter if you are in Texas or Alaska, most of us should be putting on sunscreen daily—or learning how to check the UV Index every time you go outside.

The UV Index is pretty fascinating. Our obsession with prediction is so great that scientists have instruments in the atmosphere that test the mixture of the sun's wavelengths. Based on the data collected, they can assign numerical values to the danger expected from that day's UV radiation. The scale ranges from zero to eleven (or higher), where zero is low radiation and above ten is extremely high radiation. Whenever you see a UV Index of three or higher, they recommend that you wear a sunscreen with an SPF of at least 30.

And one more note of caution. The UV Index can almost double when you are in reflective areas like water, snow, or sand. That's why people like to sunbathe on the beach instead of in their backyards, and it's also why you will sunburn a lot faster near the water. The same can be said for when you are skiing in the mountains, even though only a small portion of your face is exposed.

So, who knew: a successful beach trip not only involves the assistance of dozens of man-made polymers—spandex, insulating polyethylene, and polystyrene chief among them—but also chemical solutions we slather on our bodies to prevent waves of electromagnetic radiation from breaking up our molecules. Sunbathers rejoice!

9

PIE KID YOU NOT

In the Kitchen

When I'm in the kitchen, it's for one reason: to bake.

I love that baking is slow and methodical. It's precise. And, most importantly, it's *chemistry.* Think about it—when you bake, you need to measure everything exactly. Same as we do with substances in the lab. Then, when you're mixing everything together, you need to be super careful; you can't overbeat or under stir. Just like you can't apply too much heat or pressure when you're working with chemicals in a reaction.

The parallels continue, and in this chapter, I'm going to explain the chemistry that happens in the kitchen. Whether you're baking a pie or cooking a five-course meal, chemistry is at work.

I learned to bake from my mom, who makes killer pies. She's so good that I used to ask her to make me a rhubarb pie instead of a birthday cake every single year while growing up. In my mom's kitchen, I learned one of the most important rules of baking: to be precise.

I cannot tell you how many cookbooks I have about baking. Like my mom, pies are my specialty, so I have several books on pie crusts alone. My absolute favorite is *The Pie and Pastry Bible* by Rose Levy Beranbaum. I love how precise she is, and if you follow her instructions *exactly* on any recipe in that book, you will make the best pie of your life. But if you go rogue and change even one thing—like substituting raspberries for blueberries in her Fresh Blueberry Pie—you could end up with a sloppy, soggy mess. There's a small margin for error in baking, just like there is in the chemistry lab.

If you ask my mom, it's only possible to be a bad baker if you are sloppy with your measurements. And the easiest way to become a better baker is to invest in a kitchen scale so you can use mass (grams) instead of volume (cups, tablespoons). Not only is it faster for kitchen prep and cleanup, but a scale also helps the baker be more accurate and consistent. For instance, there's a pie recipe I use that calls for 1 ⅓ cup + 4 teaspoons of pastry flour—which is a super annoying measurement. The equivalent mass in pastry flour? 184 grams. Simple and straight to the point. When you're using a scale, you can't over heap the spoon or short the cup. 184 grams is 184 grams, no matter how you scoop it.

Now I will not buy any cookbooks that provide measurements in volume alone. Those recipes are just not accurate enough for me anymore. If you decide to start evaluating the recipes in your cookbooks, also check to make sure that they call for the proper flours for each baked good, like pastry flours for pastries and bread flour for bread. I know it might seem extreme, but flour is the most important ingredient in baking, and there's actually a substantial difference between the different kinds of flour, which we'll talk about in a bit.

Before I started writing this chapter, I looked through my pantry to see what types of flour I keep in stock, and it looks like I regularly have six different varieties: all-purpose, pastry, bread, cake, whole wheat, and gluten-free. Each one is sealed in an airtight container and properly labeled—I am a chemist after all.

In general, bakers are super finicky about their flours. There is a story about a famous British baker that was stopped by TSA upon her entrance to the States because she had an enormous amount of unlabeled white powder in her checked suitcase. She had to make a special dessert at her event, and she was not going to take any risks of having the incorrect type of flour. Unfortunately for her, TSA was not convinced that she was importing flour instead of highly restricted contraband…and that's why I always label my flour when I'm traveling!

So why are bakers so darn picky about flour? Why do we insist on using all-purpose flour for some recipes and pastry or cake flour for other desserts?

Protein. It's all about the protein.

Some flours (like unbleached all-purpose) come from hard wheats that have high protein levels, whereas other flours (like pastry flour) come from soft wheats that have lower protein levels. In fact, most of the molecules in flour are proteins.

Proteins are actually all around us—not just in food, but also in our hair and skin. As you may recall from our earlier discussion on breakfast, proteins are polypeptides, which is a fancy way of saying that the molecule is built from two or more amino acids.

There are two proteins that are found in every flour: glutenin and gliadin. When glutenin and gliadin are mixed together in a liquid, gluten is formed. Yes, *that* gluten.

Interestingly, the reason that someone with Celiac disease has to avoid gluten is that they have become intolerant to gliadin (not gluten). From a diet standpoint, it is much easier for them to avoid gluten than gliadin, hence why some people's diets are "gluten-free."

Unfortunately for those on the g-free diet, gluten is a baker's best friend (usually in conjunction with yeast). This beautiful peptide stretches to capture the carbon dioxide bubbles released by a chemical reaction with yeast, which allows for the dough to get all big and puffy. The best part about gluten is that it eventually stops stretching out and locks itself into place. This is why you always wait for the dough to double in size when baking—because that's as big as the dough is going to get once all the carbon dioxide has been released.

The percentage of gluten produced depends on the type of wheat used to create the flour. Hard wheats (called *hard* because the wheat berries are longer and harder than soft wheat berries) produce the most gluten, and therefore are the best companions for recipes with yeast. One of the hardest wheats, durum, has strong elastic properties that make it stretchy and thick, which make it an excellent flour for fresh pasta or pizza dough. It's also why hard wheat flours are found in bread pudding or other messy bread (like monkey bread).

Soft wheat berries, on the other hand, are shorter and less firm because they are filled with carbohydrates. Flours made from these berries have lower protein levels and cannot make as much gluten, so they're not going to be as stretchy or elastic. There are two primary types of soft wheat berries: white and red. Soft white wheats are perfect for pastry flours, whereas soft red wheats are better for cake flours. For this reason, I always like to use soft white wheats—or pastry flours—in my pie crusts.

One of the primary differences between pastry and cake flour is that cake flour is often bleached—or chemically treated to withstand fat and sugar more easily. Personally, I prefer the unbleached flour, so I tend to buy the King Arthur brand for all of my flours except for one. I will only use Bob's Red Mill unbleached white fine pastry flour for my pie crusts.

In addition to selecting the wrong flour, one common mistake a lot of new bakers make is to use baking soda when your recipe calls for baking powder.

Baking *soda* is the common name for sodium bicarbonate ($NaHCO_3$), which is a basic molecule that is often used in baking. (I'll talk a lot more about baking soda in the next chapter. Spoiler alert: baking soda can be used to clean your kitchen too!)

TO BLEACH OR NOT TO BLEACH

The number one question I get from bakers is, "What is bleached flour?" Is it dangerous? Are chemicals really left behind in the flour? Or is it all a marketing gimmick?

Well, bleached flour is a result of something that happened back in the 1700s. It used to be really difficult to separate the bran (the outer, darker layer of the wheat berry) and the germ (the white embryo of the wheat berry) from the flour itself. Therefore, whenever pure white flour (or pure germ flour) was collected, the millers would save it for the people in upper class. Over time, white flour became associated with wealth, which made it even more desirable.

Sketchy millers then started to add disgusting things like chalk and bone to their impure flour to make it look like white flour. This process was referred to as "whitening" or "bleaching." The English Parliament tried to pass a law banning all additives to flour back in the 1750s, but it was not enforced in any capacity.

This process still happens today, but now, only with very specific chemicals. The most common flour bleaching agent is benzoyl peroxide, which whitens the flour but does not affect the chemical integrity of the flour. Sometimes non-harmful chlorine gases are also used, but they have a very distinctive aftertaste.

But here's the crazy part: flour will naturally turn white two to four weeks after processing depending on the time of year (two weeks in the summer and four in the winter). This flour has better elasticity after the resting period, which makes it a much better dough for bakers. But most millers are too impatient to wait it out and instead prefer to use a chemical treatment.

For me, the decision is easy. Buy the unbleached flour whenever possible.

Like baking soda, baking powder also contains sodium bicarbonate, in addition to a very important acidic salt, such as tartaric acid. Baking *powder* is a chemical leavening agent, which is a molecule that increases the overall volume of the baked good. When you make a pie crust, the acid in the baking powder reacts with sodium bicarbonate (which is also in the baking powder) to form carbon dioxide gas. The gas helps to make the dough all light and fluffy, which is vital in pie making.

There are two different acids that your baking powder can contain. Fast-acting acids react with the sodium bicarbonate in the mixing bowl and immediately begin to form carbon dioxide bubbles. The two most common fast-acting acids are monocalcium phosphate and cream of tartar.

Slow-acting acids, on the other hand, need the thermal energy—or heat—from the oven to begin forming the carbon dioxide gas. Molecules like sodium acid pyrophosphate and sodium aluminum sulfate both react with sodium bicarbonate once the temperature is high enough in the oven.

I like to buy double-acting baking powder for my pie crusts, which contains one fast-acting acidic salt and one slow-acting acidic salt. My favorite baking powder is made by Clabber Girl. I like it because it contains sodium bicarbonate (the base), monocalcium phosphate (the fast-acting acid), and sodium aluminum sulfate (the slow-acting acid). It also contains a large percentage of corn starch to make sure that the powder stays dry—in other words, that the acid and base do not react before you want them to.

Another common ingredient in pie crust is butter. From a chemical standpoint, butter is considered to be a *lipid*. This term encompasses a large array of nonpolar molecules, but in the kitchen most lipids fall into a subcategory called triglycerides. When triglycerides are in the solid phase, they are referred to as *fats*. When they are in the liquid phase, they are called *oils*. For example, butter is considered to be a fat (not an oil) because it is a solid at room temperature. Olive oil is considered to be an oil because it is a liquid at room temperature.

When there is at least one double bond between two carbon atoms in a triglyceride, that's called an unsaturated fat. If there are only single bonds within the molecule, then that is called a saturated fat.

Coconut oil and butter are two common saturated (single-bond only) triglycerides. Both fats are solid at room temperature, but they are susceptible to becoming soft over time. Olive oil and canola oil are both primarily monounsaturated oils (mono meaning one double bond), but olive oil is the healthier of the two. In fact, it has a much higher monounsaturated composition compared to most of the other traditional oils.

When it comes to pie crust, I prefer good old-fashioned butter over any other lipid. But in cooking, my husband and I cycle through olive, avocado, and canola oil—and for good reason. Canola oil is a particularly good oil to use because it contains a high percentage of linoleic acid, a molecule that our bodies cannot naturally generate from other foods. (Fun fact: linoleic acid and alpha-linolenic acid are the only two fatty acids that we consider to be essential. You may already be familiar with the alpha-linolenic acid, commonly called an omega-3 fatty acid, found in walnuts and soybean oil.)

Unfortunately, there is one negative side of using these healthier oils. The double bonds in unsaturated oils like olive oil can react with the oxygen in the atmosphere, and release a stinky smell. If you've ever caught a whiff of one of your oils that made you think it's "gone bad," it probably has just oxidized and lost its double bond. That's why it's a bad idea to buy oils in bulk, unless you're cooking for a lot of people on a regular basis.

The presence of the double bond (or the lack thereof) helps us to predict the melting point of the triglyceride, where the melting point is just the temperature at which the molecule converts from a solid to a liquid.

In general, a molecule with no double bonds will have a higher melting point than a molecule with double bonds. It is usually the case that the more double bonds the triglyceride has, the lower its melting point will be. This is why most saturated triglycerides are solids (fats) and most unsaturated triglycerides are liquids (oils).

We need to know this in baking because it affects the density and flakiness of the resulting pie crust. Butter (a fat) is used to make the fluffiest pies, and in my opinion, the best tasting pie crusts. However, dough made from butter is temperamental. You have to keep it nice and cold in order to produce a kick-ass pie crust. Warm, soft butter gives you a sticky dough that is awful to work with and almost impossible to roll out—all thanks to the large number of IMFs that have formed between the dough and the rolling pin.

Some bakers prefer using vegetable shortening because it is more resistant to the IMFs formed at warmer temperatures, but

I don't think it tastes the same. These crusts are often slightly more dense than those with butter and tend to have a greasy texture. This is due to the fact that shortening is 100% fat, whereas butter is a mixture of fat (80%), and water (18%), and milk (2%).

Then there are those who like to make their lives miserable by using oil in their pie crusts. If you do this, stop it now. This resulting dough is always dry and crumbly, and mine always fall apart. Plus, oil-based pie crusts are extremely difficult to roll out, even with the best techniques and equipment.

GREASE FIRES

While I'm lecturing you on what to do (and what not to do) with oil in the kitchen, I want to give you one more point of caution. Oils/fats do not mix with water, which is why you should never use water to put out a grease fire.

A grease fire is the result of impurities in oil catching on fire. This commonly happens if you use the same oil over and over again, or if you are making a super large quantity of fried foods. The solution is to smother the fire immediately. Grab the lid to your pan (or a nearby baking sheet) and cover the fire, thereby cutting off most of the oxygen. If the fire is relatively small, you can throw a thick powder on it, like baking soda or salt. But in my household, I go straight for the oversized baking sheet.

What you should never, ever do is throw water on it. Why? Because water is polar and oil is nonpolar. This means that the two liquids are not going to mix. Instead, the much denser water is going to drop below the layer of oil, and interact with the hot pan. This is a major problem because water boils at a much lower temperature than most oils. The water will instantly vaporize and turn from the liquid phase to the gas phase. When this occurs, the newly formed gas particles will try to escape the pan—quickly. They leave the pan, pushing the layer of oil above it out of the pan and spraying the flaming oil in every direction.

The best way to avoid a grease fire is to replace your frying oil frequently and to keep a clean cooking space. So, if any oil jumps out of the pan, just use an old rag to quickly mop it up!

The last ingredient in any good pie recipe is sugar—and I'm talking about the granular stuff, not the natural sweetness that occurs in berries and fruits. Sugars are classified as carbohydrates because every sugar has carbon (*carb-*), oxygen (*-o-*), and water (*-hydrates*). Neat, right? The word *hydrate* doesn't mean that there are actual water molecules in what we call carbs. Instead, it implies that the molecule will always maintain a 2:1 ratio of hydrogen to oxygen—just like water does.

There are two main types of carbs that we interact with on a daily basis: simple and complex. Let's start with simple carbohydrates, or simple sugars. These molecules are called monosaccharides, and they are the smallest type of carbohydrate that can exist.

Two common types of monosaccharides are glucose and fructose. These monosaccharides share a molecular formula ($C_6H_{12}O_6$) but have different structures—how cool is that? When this happens in chemistry, we refer to the two molecules as *isomers*, indicating that the substances have the exact same number (and type) of atoms, but they are connected differently. For example, glucose has a six-membered ring, while fructose has a five-membered ring.

If you remember the process of photosynthesis from your freshman biology class, you probably already know that glucose is made when plants use energy from the sun to convert water and carbon dioxide into oxygen. This is one of the reasons why glucose is the most abundant monosaccharide on this planet. It's found in corn, grapes, and even our blood sugar.

Fructose, on the other hand, is the monosaccharide that is present in fruit sugars. You can find it in sugar cane, beets, honey, and of course fruits, like apples and berries.

What most of us use in our pie fillings—and in our coffee and tea—is a disaccharide called sucrose ($C_{12}H_{22}O_{11}$), or table sugar. Sucrose is not as sweet as fructose (fruit) but is definitely sweeter than glucose (found in most vegetables). What's neat about sucrose is that it's actually the result of a glucose molecule and a fructose molecule getting together. That's why it's called a disaccharide—that literally means *two sugars*. (If you guessed that monosaccharide means one sugar, you're right.)

Fructose, glucose, and sucrose all are considered to be simple sugars. These molecules can link together through condensation reactions to form a polysaccharide, or a long chain of monosaccharides. We commonly refer to polysaccharides as starches, and starches are often found in foods like potatoes, beans, and rice—things not usually found in pie.

Sugar *also* reacts to heat, like when I put my pie in the oven. This process is called caramelization, and it is usually accompanied by a change in color and an incredible array of smells. Note how when you make caramel, the white solid slowly converts to a thick yellow liquid before ultimately turning into a dark brown substance. The caramel forms when the final brown liquid hardens, which occurs after all of the fragrant compounds have been released into the atmosphere.

But what's actually happening is that the white solid you start out with, the pure sugar, is decomposing. When it interacts with heat, the bonds in sucrose break to form glucose and fructose, which is the yellow liquid that we see. Microscopically, the polysaccharide chains of monosaccharides immediately break apart into hundreds of different molecules—some sweet, some bitter, some very fragrant—which is why we can usually smell when the pie is about ready to come out of the oven.

At the same time, when I put a pie in the oven, those protein molecules we talked about earlier (like the molecules in my pastry flour) are also exposed to heat. This causes a process called denaturation to begin, where the heat of the oven starts to break apart the bonds within the protein molecules of the flour.

An easy way to visualize this is to picture the tight swirl of a warm, gooey cinnamon roll. As the oven starts to heat up, the roll begins to vibrate. The extra energy provided by the heat breaks the IMFs that held the roll together in its spiral shape and the roll begins to open up, like if you were to uncurl the roll into a long, delicious, doughy line. This is happening at the molecular level, all over the pie. In that process, what were three-dimensional proteins turned into completely flat, two-dimensional proteins (just like we saw with the eggs in our morning omelet). That matters because it exposes all of the atoms in the molecules.

Did you ever pull apart a cinnamon roll before eating it when you were a kid? Well, if you played with your food like I did, you would see where a baker has dusted it with cinnamon and butter (or frosting, my favorite) before creating the swirl. That's what the molecules in our pie flour look like after the denaturing process is done. Long lines of deliciousness at the atomic level.

Once the proteins fully denature, the next step in baking is a process called coagulation. Basically, the proteins that look like the (unrolled) cinnamon rolls begin to bump into each other. Remember that the heat from the oven is making the molecules vibrate, so it's pretty easy to bump into each other like bumper cars at a fair.

These collisions allow for hydrogen bonds and ion interactions to form. This creates a long chain of connected atoms with empty pockets between each of the big proteins. The best part of this process, to me, is that any water molecules present in the pie will jump into those pockets. This water/protein-chain combination appears on the macro scale as a "baked pie" and has a thick, crumbly texture that most people would identify as pie crust.

As bakers, we don't know the exact moment any of these molecular interactions occur. Since all of this is happening at the microscopic level, we can't look in the oven and just tell what's happening. Cookbooks never mention denaturation and coagulation. Instead, it just says to heat the oven to 350°F and bake for fifty minutes. That's why baking can be frustrating because it's super easy to undercook or overcook a dessert.

For instance, have you ever made (or eaten!) a cake that tasted dense and dry, but that had a soggy bottom? (Or watched with horror as your favorite contestant on *The Great British Baking Show* had that happen?) Soggy bottoms happen when the pie isn't taken out of the oven early enough—not from leaving your dessert in the pan for too long.

Soggy bottoms happen because the proteins in our pie have denatured and created those cool pockets for the excess water to jump into. And those proteins have coagulated too. But because our baker is distracted or forgot her timer, the pie is left in the oven minutes or even seconds too long, which results in more IMFs happening than what's needed. When a dessert gets more heat than it needs, the distance between the proteins decreases, and the proteins basically squeeze the water out of the pockets.

When water leaves your dessert, two things can happen. The first one is fairly obvious; the water evaporates up and out, leaving you with a dry dessert—or worse yet, a burned one. The second one is more surprising and frustrates lots of bakers, including me.

Since water is a relatively dense molecule, it can sink to the bottom of the pan to form hydrogen bonds with other water molecules at the base of the pan, instead of staying in those cool little pockets. If enough molecules drop to the bottom of the dessert, you can end up with a soupy layer at the very bottom that Paul Hollywood, a judge on *The Great British Baking Show*, would smirk about.

It's worth mentioning that sometimes a soggy bottom isn't the result of IMFs at all. Instead, a baker might make a seemingly innocent substitution that adds too much water to the recipe as a whole. For example, if a recipe calls for four cups of raspberries, you probably shouldn't substitute four cups of blackberries even if you prefer that flavor. Here's why: a raspberry has a lot less natural water content than a blackberry does. That's why adding in fresh, juicy blackberries instead of less water-dense raspberries often results with a sloppy, soggy mess instead of a pristine pie.

FRESH VERSUS FROZEN BERRIES

Have you ever wondered why some recipes call for frozen fruit, while others want you to use fresh? That's because there's a major difference between the blueberries in your freezer and the blueberries in our fridge: the length of the hydrogen bonds that exist between the water molecules. The hydrogen bonds in the liquid phase require *less* space than the hydrogen bonds in the solid phase. Water is quite unusual in this manner—the large distance between the molecules in the solid allows for solid ice to float on top of liquid water. This is the exact opposite of most solids/liquids. The majority of solids will sink in their liquid.

What this means is that when water freezes, it expands (most solids shrink). This is why you cannot leave a bottle of champagne in the freezer. The water freezes, expands, and pushes the cork out of the bottle creating a massive explosion in the freezer. Even as a chemistry major, I had to learn the hard way that the same thing happens to beer bottles. That was an embarrassing mistake.

But now, let's consider how this science affects the taste of fresh versus frozen berries. Fresh blueberries have a standard water content (approximately 85% water), giving it the perfect juicy:crunchy texture. Frozen fruits, however, do not. When blueberries are placed in the freezer, the internal water freezes. The newly formed ice pushes against the edges of the cell membranes, sometimes damaging the cell if not rupturing it altogether.

When the berries are removed from the freezer, the ice melts to leave behind mutilated cell membranes. This change affects the ability of the berry to retain the water, which ultimately affects the overall water content in the pie (and the taste of the dessert). In other words, if the pie calls for frozen berries, make sure you use frozen berries, in order to avoid making a soggy-bottomed pie.

When the recipe is followed exactly, a perfectly cooked pie can make your kitchen smell like heaven. The molecules that we smell are called aromatic compounds, and tons of them are released when we pull a pie out of the oven.

In most cases, the smell of food is directly correlated to how that food tastes. A "good-smelling" food usually tastes really good, and it can even trigger memories. I know I get a wave of nostalgia when I smell one of my mom's pie recipes baking in my oven. That familiar smell provokes our memory and affects how we perceive the taste of food.

Our sense of smell is our first line of defense in the kitchen. Its primary purpose is to keep us away from things that could potentially kill us, like bacteria. A small percentage of people do not have a sense of smell, and not only do they not get to experience the full spectrum of tasting food, but they also don't have the human instinct that protects us from eating rotten or spoiled food. I actually know someone who lacks a sense of smell. His mom visited him in college once and nearly vomited the second she stepped into his apartment. Apparently, he had some bad chicken in a hidden corner of his fridge, but couldn't smell it.

For the rest of us, however, when a meal smells good *and* tastes good, the two senses have combined to form what is referred to as flavor. The flavor of the meal is what you respond to—and each of us has a particular set of flavors that we enjoy the most. That said, every flavor in the universe, from Kraft macaroni and cheese to the tasting menu at a fancy restaurant, is derived from four molecules: water, fats/oils, proteins, and carbohydrates.

Our brains are really good at deciphering between these tastes down to a microscopic level. In fact, our brain can even detect whether we're eating a monosaccharide or a polysaccharide (a.k.a. a sugar or a starch). That's because our taste buds, which send messages to our brain, can detect tons of different molecules. For example, when a taste bud identifies hydrogen ions (H^+), we perceive the food to be sour. Alkali metals, on the other hand, make things taste salty.

What's important about this when it comes to baking is that our brain can detect the difference between a monosaccharide—the sugar in the fruit mixture—and a polysaccharide—the starch in pastry flour. I would argue that it's the mix of sweet (monosaccharide) and savory (polysaccharide) that makes pies the best dessert ever. (I might be biased—did I mention that my mom makes killer pies?)

Our taste buds are able to detect all these molecules because our brains closely monitor the concentration of specific ions—in this context, Na^+ and H^+—within what's called our ion channels. These ion channels are located within the cells in our organs, and they provide distinct pathways for the ions to move throughout our body, just like roads are made for cars to move from one location to another.

When we take a bite of something that has a lot of salt, our brain detects an increase in the number of sodium ions moving within an ion channel on our tongue. When the concentration of hydronium ions increases, our brains immediately know we have eaten something sour.

And all of that happens in an instant. Our brains are really powerful.

From a molecular perspective, there's one really big difference between salty/sour and sweet/savory—the bonds between the molecules. Salty and sour foods use ionic bonds whereas sweet and savory foods use covalent bonds. That's why we can tolerate really sweet foods but not super sour foods. For example, when we eat blueberry pie, our taste buds immediately detect the sweet taste. But since we are eating something sweet, we're not using the ion channel pathways.

In the same way, bitter tastes are consistent because their concentration does not change the overall taste. It does not matter if you drink one drop or one cup of the food, the taste is equally bitter.

Because sweet, savory, and bitter tastes don't travel to the brain using our ion channels, they're always put in the same category. These tastes come from a chemical reaction between specific covalent molecules and the receptors in the cell membranes of your taste buds. The second this reaction happens, our brains detect the taste of sweet, savory, or bitter. Again, all of this happens in less than one second.

And while we're here, I want to quickly clear up a misconception that I frequently hear on this topic. Your whole tongue can detect all five tastes relatively equally. It does not have different sections of taste buds! Every inch of your tongue can detect the sweetness of your pie.

Overall, there are five main tastes in food: sweet, salty, sour, umami, and bitter. (The word *umami* is a Japanese word that directly translates to *deliciousness*. However, most people use the word *savory* instead.) Great bakers use these five categories to build endless combinations of amazing flavors.

Just look at a classic rhubarb pie. The filling contains 4 cups of rhubarb (sour), 2/3 cup sugar (sweet), and a pinch of salt. When combined with some lemon zest (more sour), the perfect balance of salty-sweet-sour deliciousness is created.

But what I find to be especially interesting, from a chemistry perspective, is that the same combination of molecules can be interpreted by each of us differently. Some people hate rhubarb pies, yet I can't get enough. Why is that?

Flavor preferences are all based on the psychology of pleasure, which helps to explain why people have favorite foods, in addition to favorite colors, movies, songs, etc. While brain chemistry is extremely complicated, most psychologists generally agree on one theory: people develop a favorite thing based on a positive experience they had when they were first exposed to it…and their brains respond to different chemical receptors as a result.

When it comes to food, most of our favorites were identified when we were very young. I likely developed my love for rhubarb pies because it was the first pie that I ever had. The sweet-tart-salty mixture blew my young mind, and I have never had another pie that has beaten that single experience.

But there is one exception to this general theory: you can actually train your tongue to detect more flavors. Just like you can train your muscles for a marathon or a football game, with hard work, dedication, and a lot of exposure, you can learn to detect the different molecules in food. And when this happens, people tend to discover new foods that they love all because they have refined their palate—a fancy way of saying they've increased the number of flavors they can detect.

Some people have excellent palates. For example, I've met bakers who can instantly recognize a hint of nutmeg in oatmeal cookies or a foodie that can figure out that their favorite Thai place uses fish sauce in a particular curry. For most of us though, the older you are (or the more you smoke), the more difficult it is for your brain to interpret the signals from your tongue. It's almost like your taste buds—or your ability to detect ionic and covalently bonded molecules—get worn out or lazy, especially as you reach old age. So, while you are young, get out there and try different things. Make a rhubarb pie and then an apple pie, and see which one you like better.

Hopefully, now that you know what is happening between the atoms and molecules in your dessert, it will make the entire process of baking a lot more fun…and eating a lot more interesting.

But if you are anything like me, you just destroyed your kitchen while baking these incredible blueberry pies. Your clothing (and hair) is covered in flour, and your dogs are having a field day licking dessert fragments off the floor.

As my pie enters its four-hour cooling period, I feel the urge to walk straight into my laundry room to collect a pile of old rags and an armful of cleaning products.

I have some work to do.

10

WHISTLE WHILE YOU WORK

Cleaning the House

I like to clean.

Actually, that's not entirely true. I like how it feels when my house is clean. Sometimes, after I make something look all sparkly and new, I force my husband to get up and admire whatever thing I just cleaned. Over the years, he has learned to look at the toilet and just say, "Oh yes, very clean," before going about his day.

And of course, a part of me enjoys that any time I bleach my countertops or use a lemon to unclog a drain, I get to put my chemistry skills to work at home.

But before I ask you to read an entire chapter about disinfectants—and hopefully give you a few tricks and hacks along the way—I want to explain to you why you should care about the chemicals you use every time you do a little cleaning.

To start with, every household cleaner contains an array of carefully selected molecules that can work together to perform a specific cleaning task. Companies use acid in toilet bowl cleaners, sodium hypochlorite in bleach, and ammonia in window-cleaning products. Each of these molecules is very good at removing the gunk in their designated location, but they can be harmful to other surfaces. I'm sure you already know this intuitively, since most people have no interest in using the shower cleaner to mop their floors, or Windex on granite counters (which will strip the protective coating off them).

More importantly, these chemicals should never be combined to form a more "powerful" cleaner. That would be like me walking into a lab and combining random molecules just to see what would happen—except worse, because the chemicals used in cleaners are designed to be reactive. Case in point, toilet bowl cleaner and bleach.

When a strong acid (toilet bowl cleaner) is added to sodium hypochlorite (bleach), the resulting chemical reaction can produce a toxic gas called chlorine. Chlorine gas, sometimes called bertholite, was used as a chemical weapon in World War I. Although I have never smelled it, the soldiers in World War I described the gas as having a distinct pineapple-and-pepper smell. The gas reacts with the water in your mouth, throat, and lungs to form hydrochloric acid. Bertholite is a nasty molecule that you definitely do not want to accidentally produce in your kitchen or bathroom—or any small space for that matter.

You also do not want to combine bleach with any products that contain ammonia (like window cleaner). When sodium hypochlorite and ammonia combine, they react to form a few different chloramines (NH_2Cl), which are believed to be unhealthy for us. There have been a few studies that have linked bladder and colon cancer to communities that have a higher concentration of chloramines in their public water and/or swimming pools, and they have been shown to cause eye irritation and respiratory issues.

But just in case you need one more horror story to refrain from becoming a cleaning product chemist, in 2008, a woman in Japan decided to mix laundry detergent with another cleaner, ultimately killing herself and hurting ninety other people in her apartment complex. Due to public safety concerns, the media in Japan decided not to report the name of the other cleaner, which I think was a wise decision.

So, now that we know to use one cleaning product at a time, let's take a look at the powerful science behind cleaning gunk, goo, stains, and messes. And along the way we can ponder whether we have in fact become *too good* at cleaning our houses.

Let's start in the kitchen because that is where I begin my Saturday morning cleaning, and every Biberdorf cleaning tour. The first thing I do is gather all the dishes from the night before and load the dishwasher as much as possible. Plastics go on the top because the heat from the dishwasher can alter the shape of the container (chemistry!), and big pots and pans go in the bottom.

The science of the dishwasher is pretty simple. Water flows into the machine and then the dishwasher detergent is released. It is imperative that you do not confuse dish soap with dishwasher detergent because they are comprised of two completely different types of molecules. Dish soap is made of molecules that are safe for your skin, while dishwashing detergents use harsher chemicals that you would never, ever want to put directly on your skin.

The strong molecules in the dishwashing detergent eat away at the gunk on your plates and silverware before being sucked down the drain. Metasilicates, sodium carbonate, and metal hydroxides are present in most dishwasher detergents—many of them are even combined with enzymes. Once the dishwasher is started, these chemicals react with the molecules on your plates in a variety of different ways. The alkaline salts dissolve the grease on your plates, while the enzymes simultaneously go to work on the protein fragments. If neither of these molecules can react with the caked lasagna on the pot, then the metal hydroxides will finish the job.

And in tandem, these chemicals loosen all the bits and scraps on the plates, which get broken down further in the hot, chemical stew. After this, they travel down the dishwasher drain before those dishes are rinsed clean. Voilà!

Here's a fun and fascinating story about the time I was introduced to "dishwasher chemistry" in my sophomore year of college, when I noticed that bubbles were pouring out of my dishwasher at an alarming rate. Turns out, my least domestic roommate had filled the small container meant for dishwashing detergent with dish *soap*. To top it off, she wanted to make sure that the dishes were extra clean, so she squirted more dish soap on top of every single dish.

I am not exaggerating when I tell you that we had bubbles pouring out of our dishwasher for days. We finally broke down and called maintenance, and a repairman performed a real neat science trick: he arrived at our door with a large container of vegetable oil, walked right over to our dishwasher and dumped at least a cup of oil into it, before telling us to run the dishwasher twice, and then left.

The results were instantaneous.

The bubbles stopped immediately because of the way the oil reacted with the surfactants within the dish soap. These big surfactant molecules have a hydrophilic side and a hydrophobic side that are traditionally used to help remove the grime from the dishes that you wash by hand. The hydrophobic side grabs the food fragments and the hydrophilic side attaches to the water, allowing for the food to glide off the dinner plates with ease (just like the surfactants in our shampoos remove the grease from our hair).

But when our hero maintenance man added the oil to the dishwasher, the hydrophilic side of the surfactant formed hydrogen bonds with the water, while the hydrophobic side of the surfactant formed new dispersion forces with the oil. The water was then flushed out of the dishwasher and pulled the oil molecules out too.

What was the reason for the bubbles in the first place? The bubbles were produced when the surfactant in the soap formed hydrogen bonds with other surfactant molecules (yes, from the same dish soap) or other water molecules. These interactions were so strong that it trapped air bubbles that were formed within the dishwasher. Millions of them; hence our bubble problem. But when the oil was added to the dishwasher, the hydrophobic side of the surfactant was activated, which ultimately broke up the bubblefest in the dishwasher.

This is also the reason why dish soap is so great at removing the grease (oil) from your pots and pans: the dual hydrophilic/hydrophobic qualities pull the food and grease particles away from their previous bonds (with your pans). And why it is usually easier to add dish soap directly to a greasy pan instead of trying to clean the pan in dishwater. The oil in the pan repels the water in the sink, therefore we need a middleman—the surfactant—to pull the oil off the pan before it can be flushed away with water.

Be careful to avoid using dish soap on a cast iron pan though. A quality cast iron pan has been seasoned to have a thin layer of molecules that covers every inch of the bottom of the pan. If you use dish soap on the skillet, the hydrophobic side will bond with the molecules on the pan and rip them off the surface.

According to my girl Rachael Ray, the best way to clean the pan is to use extremely hot water and kosher salt: if you rub it in, the corners of the salt crystals work their way into the offending molecules and physically push them off the cast iron surface, but without reacting with the seasoning molecules. Use hot water to rinse the skillet before adding a tiny layer of oil to the bottom of the pan. She advises you then to cover it with a paper towel to prevent the formation of rust (but I always skip this step; the oil will repel any water in the atmosphere, therefore the paper towel is not really necessary).

While dish soap surfactant works great for pans, it is completely useless when combatting the dark stains that your plastic Tupperware collects. For that task, I turn to my trusty companion, sodium bicarbonate ($NaHCO_3$), commonly referred to as baking soda. I don't know about you, but I have baking soda all over my house—one box to use in our kitty litter, one to use in my science experiments, and one to use in my pies. This one small molecule can do so many different things, just because it is a base.

Molecules that are bases (like baking soda, or sodium hydroxide) feel slippery and slimy to us because they react with the fats and oils on our skin. They literally feel slippery *because* you are touching them, and they are drawing out the oil on your fingers. Gross, right? If you have enough skin-to-molecule contact, some bases can actually form soap, right on your skin.

Bases gained a little notoriety back in 1999 in the movie *Fight Club* when Tyler Durden (Brad Pitt) poured a base, the aforementioned sodium hydroxide, on Edward Norton's hand. Norton freaks out and screams in pain as the base reacts with his skin. The science is wrong in this part of the movie—sodium hydroxide on your hand doesn't hurt *that* much—but it is super icky when it starts to form soap from the fats and oils on his hand. You may have experienced this feeling any time you have used a great deal of any household cleaning agent that's a base, like baking soda.

In order to explain this, and how sodium bicarbonate removes the stains from the Tupperware, it's important to understand what a base is. Bases are often defined as molecules that accept a proton (H^+) when added to water. In this context, the word proton is just the word scientists like me use to describe a hydrogen atom that has lost one electron. In the case of baking soda, the sodium bicarbonate will "accept a proton," which looks like this:

$$NaHCO_3 + H^+ \rightarrow Na^+ + CO_2 + H_2O$$

For the sake of our discussion, we just care that the sodium bicarbonate will accept a proton from the molecules that are staining the Tupperware. This process takes a little while so I usually recommend soaking the stained plastic in a solution of aqueous baking soda for a few hours. At the end of the soak, I also add a few squirts of dish soap to add a little surfactant to the mixture.

The baking soda removes the stain by stealing a few protons and forcing the molecule to decompose before the dish soap flushes out the offending molecules (thanks to the surfactant molecules). Some people add ice to the concoction, but all that does is decrease the amount of baking soda that dissolves into the water, which is counterintuitive.

On a microscopic level, all bases want to accept protons (H^+), but the fastest and easiest way is for them to take the protons from other molecules called acids—molecules that are very reactive and have an extra proton at the ready. A great example of an acid is vinegar, which contains about 5% acetic acid (CH_3COOH). The reaction between baking soda and vinegar is delightful—the stuff thousands of science fair volcanos were made of.

Here's how it works: when you pour vinegar over baking soda, the acetic acid (CH_3COOH) in vinegar donates its proton (H^+) to the sodium bicarbonate ($NaHCO_3$), as shown below:

$$CH_3COOH + NaHCO_3 \rightarrow CH_3COONa + CO_2 + H_2O$$

Immediately, the mixture starts to bubble. These bubbles are just the carbon dioxide gas that is produced during the neutralization reaction.

But there is another reaction happening while the bubbles are being released. When the acetic acid (CH_3COOH) donates its proton, it becomes sodium acetate (CH_3COONa). In acid-base chemistry, these molecules are referred to as conjugate acid-base pairs because their molecular formulas only differ by one proton. Acetic acid (vinegar) is the acid, and sodium acetate is its conjugate base.

Luckily for us, sodium acetate isn't too bad for us, so the combination of vinegar and baking soda is considered to be completely safe.

And it won't surprise you to learn that vinegar, especially white vinegar, is another standard household cleaner that can be a lifesaver in the kitchen. (You can use other vinegars too, but darker vinegars, like red wine vinegar, aren't usually recommended for cleaning lighter surfaces.)

White vinegar is a clear, colorless liquid that is quite inexpensive, and a fantastic reagent for removing the stains in your kitchen without causing any damage to the appliances themselves. It can be used in your sink, in your coffeepot, and on cloudy wineglasses. Some people even use vinegar to clean their garbage can.

When the acetic acid donates a proton to the grime in your kitchen, like your sink, the molecules basically (ha!) let go of the sink and preferentially engage in acid-base chemistry with the vinegar. This takes some time so the vinegar needs to soak on the sink surface before you will see any noticeable difference. But after fifteen minutes or so of a vinegar treatment, you can use a scrubbing brush (or an old toothbrush) to remove any grime and debris before thoroughly rinsing the entire surface with water.

Do not ever do this, but if you were to taste that vinegar-water mixture, it would have a distinct sour taste that is typical of any acid. The sour taste is a result of the high concentration of hydronium ions (H_3O^+) present in a solution. This also happens when beer is over-fermented; acetic acid forms and then dissociates to produce hydronium ions, which gives the home brew an extremely sour taste.

An acid that you *could* safely taste is the citric acid found in lemons and limes, another safe household cleaner. I like to use lemons to clean out the sink drain, in addition to my refrigerator water dispenser. In Austin, we have relatively hard water, which means that we have all kinds of minerals in our water. Over time, they collect on the inside of our pipes and can lead to clogs, especially when other items like food or hair get caught up in these edges.

Lemons—thanks to their citric acid content—can be a miracle solution for exactly this problem. Just chop one or two lemons in half and use your garbage disposal to push them down your pipes. Use warm water to flush the citric acid down the drain, and enjoy your new lemon-scented kitchen.

Citric acid is what's known as a triprotic acid. This means that it has three different protons that it can donate in acid-base chemistry. In the context of the kitchen drain, this powerful acid essentially bear-hugs the built-on hard water minerals as it moves throughout the pipe. The minerals are so attracted to the acidic molecule that they let go of the pipe, which unclogs the drain.

Are you sensing a theme here? All of the cleaners we're using in our kitchen, and often the bathroom, do a nifty job of attracting and grabbing/holding on tight to other molecules, and thereby removing them from the places we don't want them to be. But these molecules have wildly different chemical compositions, so it takes a different "magnet" every time.

Let's talk about countertops. If I do not have much time, I grab my all-purpose surface cleaner, which is primarily water with a dash of dimethylbenzyl ammonium chloride. Like dish soap, this molecule is another surfactant that has a hydrophobic side and a hydrophilic side, therefore it just needs a few seconds to initiate a physical change (form IMFs, primarily dispersion forces) with the dirt on the surface. You should always follow the directions on the cleaning products, but if you're like me, you might not be great about waiting for a few minutes when it comes to all-purpose surface cleaners.

That's why I'm a little neurotic about using bleach (sodium hypochlorite) on my kitchen surfaces at least once a week. In the form of liquid bleach, sodium hypochlorite is a pale yellow-green color that has a very distinctive smell commonly associated with cleanliness. Sodium hypochlorite is a basic molecule, therefore we would expect for a solution of bleach to behave in a similar way to sodium bicarbonate (baking soda).

Each cleaning solution that contains bleach has a different concentration of sodium hypochlorite than the others. Laundry detergents and common household bleaches are usually 3–8% sodium hypochlorite, but they also tend to have a little bit of sodium hydroxide in them too (the base used in *Fight Club*).

The sodium hydroxide is not added as an extra cleaning bonus. Instead, it is used as an emergency safety feature that helps to slow down the decomposition of sodium hypochlorite. If, during the storage process, the bleach decomposes to release that toxic chlorine gas I mentioned earlier, the sodium hydroxide will then react with the gas to reproduce more sodium hypochlorite. Neat, huh?

I began using bleach in my kitchen in college after a chemistry professor went on a tangent about the wonders of sodium hypochlorite. He expressed that it is a favored disinfectant in hospitals because it can safely kill the microbes on a bunch of different surfaces. At low concentrations—like 0.05%—sodium hypochlorite can be used to disinfect a doctor's hands. At higher concentrations—like 0.5%—sodium hypochlorite can actually disinfect surfaces with bodily fluids on them. This is why bleach is the number one chemical used on blood spills.

This being said, bleach does not actually *remove* the foreign molecule from the surface. Instead, it breaks a few of the bonds within the molecule (which kills the bacteria), but the building blocks—or the atoms—are still clinging to the pores of the countertop or the bathroom floor.

So if you're trying to hide something from the police, you shouldn't use bleach.

Let me explain. The sodium hypochlorite reacts with the molecule and changes the way it interacts with light. After the reaction, the molecule can no longer emit light in the visible region (red-orange-yellow-green-blue-indigo-violet). This means that the offending molecule is now invisible to the human eye, but it is *still there.*

Therefore, if you have a massive blood spill that you are trying to hide from the police, you can use bleach to remove all visible traces of the blood. However, if they are suspicious of your behavior, all they have to do is throw some luminol on the bleached area to encourage the blood to light up like a firefly under a black light (UV light).

In other words, if you use bleach on your countertop or in

your shower, you are not actually scrubbing the bacteria off the surface. Instead, you are just changing the color of the molecules. Don't worry though, most bleaches contain a small amount of surfactant to help you remove the bacteria with a wet rag.

After I finish in my kitchen, I typically move into my living room, where I get tremendous satisfaction from removing the dog nose prints from my reading-nook window.

The best candidate for this job is another base called ammonia. Since it's a base, ammonia binds to dirt and grime, which makes it an excellent cleaning product. With just one wipe, I can remove the dirt from the glass window pane, giving me a squeaky-clean reading-nook window that my dogs proceed to ruin as soon as I turn my back.

Ammonia is also great for polishing furniture or floors, but it will not do much in a toilet bowl or on shower grime. But don't tell that to the dad in *My Big Fat Greek Wedding* because he believed Windex could cure anything—even acne.

Since I have two dogs, a cat, and a husband with allergies, the other living room chores mostly consist of dusting and vacuuming up animal hair. Hence, my obsession with my Swiffer and my Roomba.

Swiffers are serious business. Isn't it amazing how simply running a unique piece of fabric over horizontal surfaces eliminates a large amount of particulate matter from any living room? The next time you pull out your duster, take a minute to examine it. Look at the amount of surface area it has, and inspect the way the fibers are entangled with each other.

Then, while you use it, watch how the small fibers collect the dust with each passing sweep. All you are doing is allowing for the IMFs to form between the dust particles and the Swiffer cloth. Some people refer to this phenomenon as static cling, but I just call it chemistry.

My Roomba—named Stevie—doesn't use any physical or chemical changes to pull the dust off the floor. Instead, he uses a motor to spin a fan to create a "vacuum" (really just a low pressure situation) to suck the air molecules and accompanying dust into the robot. The air is filtered out the other side of the machine, but the dust and animal hair is collected within the robot itself.

Of course, we all know we should dust first, and then vacuum. But here's a tip: if you have some time, wait for a few minutes before vacuuming after dusting. Most particulate matter is light enough that it can hover on the gaseous molecules (nitrogen, oxygen, argon, carbon dioxide) in the air, and they need a few minutes to settle down on your floors.

The most chemical-laden room in the house, is of course, the bathroom. When I do my bathroom cleaning, I instinctively want to put on my goggles and gloves since I know that I will be working with strong acids and bases. They are so much stronger than the molecules we use in the kitchen, especially if you use one of the most powerful cleaners of them all, Drano.

In liquid form, Drano contains sodium hydroxide (lye) and sodium hypochlorite (bleach). Drano is a very strong base and should never, ever be put into your body because it is an extremely corrosive material. If you ever do accidentally get Drano on your skin, stop what you are doing immediately and flush the body part with water for ten minutes.

What makes baking soda and Drano so different chemically? They are both bases that are used as cleaning products. However, one can safely be used in our blueberry pies, while the other can kill you if ingested.

Both molecules are bases, which means they will act similarly in water. But sodium hydroxide is a strong base and sodium bicarbonate is a weak base. Huge difference.

If a base is strong, all the reactants will be converted into products. If it is weak, only some of the reactants will be converted into products. While this may not sound like a big deal, this distinction is crucial in determining how effective cleaning supplies will be. So, how can we tell if our base is strong or weak?

SODIUM HYDROXIDE IN THE KITCHEN

You might also find sodium hydroxide in your kitchen pantry, in the form of food-grade lye. I highly recommend the products made by the Modernist Pantry. They have lots of safe food-grade options that are very fun to use in the kitchen.

My favorite thing to make with lye is tiny pretzel bites. If you are looking for a good recipe for pretzels, check out Alton Brown's recipe for Homemade Soft Pretzels. For those of you unfamiliar with his work, he's another nerd that loves to talk about the science in cooking, and I can't say enough good things about his show *Good Eats*.

Anyway, there are two different ways you can make pretzels: use sodium hydroxide (which you can easily find in the form of food-grade lye) or baking soda. For both methods, the dough is added to a solution of gently boiling base, which causes the outside of the dough to turn a light yellow-brown color. The base breaks down the long polypeptide chains found in the flours of the dough.

The reaction with the base produces smaller amino acids that are active in the Maillard reaction—the chemical reaction that gives Bavarian pretzels their distinctive brown color and savory flavor. In order for this reaction to occur, one amino acid has to react with one carbohydrate. Since fructose and glucose are smaller than sucrose, they tend to initialize these chemical reactions by grabbing a terminal atom on the amino acid.

When in the oven, the heat encourages the decomposition of the exterior molecules to occur, which produces hundreds of different molecules. The majority of these new molecules are also brown (like in caramelization), but the resulting flavors are different. Since the Maillard reaction includes amino acids (proteins) instead of just carbohydrates (sugars), the Maillard flavors are often described as "meatier." The nitrogen atoms from the amino acid provide a much more complex flavor than what's produced by caramelization.

Since baking soda is a weaker base than sodium hydroxide, the chemical reaction between the baking soda and the proteins in the flour doesn't react to the same degree as the sodium hydroxide. Fewer amino acids will be activated for the Maillard reaction, and therefore the baking soda pretzels are usually not as dark.

We can evaluate how much a base dissociates (or breaks apart) by using something called potential hydrogen. You might know it as pH or the pH scale, which ranges from about 0 to 14. The pH scale is a logarithmic scale that chemists use to determine if a product is basic or acidic. Once we know if a molecule is an acid or a base, we can predict *how* the molecule is going to interact with other molecules. In the context of cleaning supplies, the chemical properties of the acid or base will determine where the household cleaner can be used—in the kitchen or the bathroom.

Neutral species like pure water have a pH of 7. Bases always have a pH *above* 7 and acids will always have a pH *below* 7. In order to measure the pH of a solution, we can use a pH probe or a piece of pH paper. The probe can be dipped into the solution, and it will spit out a number. The pH paper, which is a significantly cheaper method, changes color. You can then use a pH scale that matches different colors to the numbers 0–14.

But what is the pH probe or pH meter actually detecting? They are measuring the concentration of hydronium ions (H_3O^+) and hydroxide ions (OH^-) in the solution. A pH greater than 7 means that there are more hydroxide ions (OH^-) than hydronium ions (H_3O^+), and these solutions are basic. Some common examples of bases are shampoos, saltwater lakes, and the majority of our cleaning supplies.

Like we discussed earlier, depending on your location, you may even have basic (or "hard") water. There is some research that suggests that drinking basic water helps to minimize the frequency of heartburn. Brands like Essentia and AQUAhydrate have lots of hydroxide ions but it's not because they naturally are more basic. These companies intentionally add minerals to the bottled water to increase the pH. Personally, I cannot stand the taste of basic water, therefore I tend to favor bottled waters that are more acidic, like Dasani and Aquafina. But just so we're clear, the pH of the water you drink does not matter at all once it's inside your stomach...it's just a matter of taste!

Where it does matter, of course, is in our chemical cleaners. Strong bases, like the sodium hydroxide in Drano, have *very* high pH values around 13 or 14 because they have an extremely high concentration of hydroxide ions in the cleaner. The more hydroxide ions in the solution, the higher the pH of the solution, and the more corrosive the substance. Drano operates just like the lemon does in the kitchen drain, except it is much more powerful. Instead of bear-hugging the minerals and then moving throughout the pipe, the sodium hydroxide rips the minerals off the pipes like a heartbroken teenager rips pictures of her ex out of her locker.

Baking soda and ammonia are both weaker bases than sodium hydroxide. Therefore, they have lower pH values than Drano, but higher values than neutral water. Baking soda has a pH of 9, whereas ammonia is closer to 11. Since the pH scale is a logarithmic scale, there is still quite a big difference between the number of hydroxide ions in a solution of ammonia versus a solution of sodium hydroxide.

When there are more hydronium ions than hydroxide ions in a solution, it is considered to be acidic. This means that the liquid will have a pH lower than 7, just like vinegar, fruit juices, and tomatoes all do. Acetic acid and citric acid are both weak acids with pH values around 3.

Strong acids, like the hydrochloric acid commonly found in many toilet bowl cleaners, have pH values closer to 0 or 1. At local hardware stores, it is usually just called Acid Bowl Cleaner, but the science is the same as the other acids I mentioned earlier. The hydrochloric acid attacks the stains (and bacteria) found in dirty toilet bowls. It dissolves the grime and gunk by ripping apart the bonds within the molecules, after which the molecular fragments can be easily flushed down the toilet.

But what happens if you take something that is very acidic and add it to something that is very basic? For example, let's pretend we decided to do something stupid and mix toilet bowl cleaner (the super acidic hydrochloric acid) with Drano (the basic sodium hydroxide)? When these two molecules combine, a neutralization reaction occurs. This type of reaction will happen whenever an acid and a base are near each other. The neutralization reaction earned its name because the pH of the final solution is often close to 7, which is neutral. For our example, we have a strong acid and a strong base reacting with each other to form saltwater, like this:

$$HCl + NaOH \rightarrow NaCl + H_2O$$

But now, let's remove the spectator ions—aptly named for their lack of action in the reaction—from the chemical equation. When we do that for the neutralization reaction between a strong acid and a strong base, we get the below generic equation:

$$H^+ + OH^- \rightarrow H_2O$$

Look familiar? When we mix a strong acid and strong base together, they cancel each other out to give us a resulting solution of mostly water and a little salt (NaCl). By looking only at this chemical equation, I can see how one could think that mixing toilet bowl cleaner and Drano together would not be that bad. And in this instance, you are mostly right.

The problem is there are smaller amounts of other active chemicals in each of the cleaners that cannot be mixed together. For most neutralization reactions, the resulting solution is not just saltwater. In fact, there are usually two (or more) products in the chemical reaction between a weak acid and a weak base. Think about the classic vinegar + baking soda = volcano reaction, which we mentioned earlier. Adding stronger bases and acids can make this process dangerous indeed.

Now, it may surprise you to learn that there are actually some products out there, called buffers, which are solutions that are *supposed* to contain an acid and a base. These are not some random concoction that you whip up in your kitchen; buffers are made by mixing a weak base and its conjugate acid, or a weak acid and its conjugate base. Remember, these molecules are referred to as conjugate acid-base pairs because their molecular formulas only differ by one proton. Right now, you have a few natural buffers in your body, such as the phosphate buffer that maintains the pH in your kidneys and urine.

BODY BUFFERS

Buffers also play a very crucial role in the regulation of the pH in our blood. For this to happen, our bodies take the carbon dioxide we produce during respiration and react it with water to produce carbonic acid (H_2CO_3), like this:

$$CO_2 + H_2O \rightleftharpoons H_2CO_3$$

Once carbonic acid forms, it releases a proton to form the bicarbonate ion (HCO_3^-), like this:

$$H_2CO_3 \rightleftharpoons H^+ + HCO_3^-$$

This reaction, in combination with the previous reaction, is the basis of the carbonic acid–bicarbonate buffer system, which maintains a pH of 7.4 in our blood.

For example, if you do something that happens to increase the concentration of the hydronium ions (H^+) in your blood (like when you work out), the pH will naturally decrease (remember that acids have low pH values). When this happens, carbonic acid is formed momentarily before decomposing back into carbon dioxide and water, like this:

$$H_2CO_3 \rightleftharpoons CO_2 + H_2O$$

The carbon dioxide is then pushed out of the capillaries into the air space of the lungs, where it can easily be exhaled. This entire process brings the pH of the blood back up to 7.4.

If the pH of the blood gets too high, this means that there are too many bicarbonate ions in the blood plasma. The bicarbonate ion is a conjugate base, which means it will have a higher pH. In these cases, our bodies naturally react by changing their breathing rate to push the gaseous carbon dioxide out of the lungs and into the bloodstream. Here, it can be quickly converted into carbonic acid, which lowers the pH back to healthy levels.

Buffer solutions are powerful tools in the laboratory because they resist minor changes to pH. For the exact same reason, they are ideal cleaning products for your swimming pool and hot tub—two things I hope to have someday. For the sake of this discussion, let's all pretend that we have access to a private swimming pool and a glorious hot tub. In this case, we would all likely use a buffer to kill the bacteria and microorganisms in the water of our pools/hot tubs, without affecting the pH of the water itself. The most common buffer is the solution made from hypochlorous acid and the hypochlorite ion (i.e., a weak acid and its conjugate base).

The perfect hypochlorite-hypochlorous acid buffer will have 50% weak acid and 50% conjugate base—or 50% hypochlorous acid (HOCl) and 50% hypochlorite ion (OCl^-). When made properly, this buffer will maintain a pH of 7.52. That means that it will be able to resist large changes in the pH when small amounts of acid or base are added to the buffer.

Let's look more closely, using the hot tub as our example. If a foreign object that is slightly acidic (like a mimosa) accidentally falls into the buffered water, the base component of the buffer will react with the acid to neutralize the "threat." In this case, that means that the hypochlorite ions (OCl^-) will react with the hydronium ions (H_3O^+) in the acid (mimosa). After all of the acid has been neutralized, we would expect for the pH to drop a little bit, but still remain close to 7.5.

But if a foreign object that is slightly basic (like hand soap) is added to the hot tub, the base component of the buffer would not be able to do anything to minimize the changes in the pH. Instead, the acidic part of the buffer would kick in and react with the base to neutralize it. In this case, the hypochlorous acid (HOCl) would react with the hydroxide ions (OH^-) from the base (hand soap). Again, we would expect the pH to change a tiny bit during this process, but for the pH of the final water to still be around 7.5.

Now let's pretend that a saboteur has jumped into your backyard and dumped a ton of bleach into your hot tub. If that happened, the hypochlorous acid in the hot tub would continue to react with the sodium hypochlorite until all of the hypochlorous acid has been used up. When this happens, the pH will begin to increase drastically from 7.52 to values around 12 or above.

On the other hand, if this jerk decides to add a bunch of battery acid to your hot tub, the hypochlorite ion will react with the acid until all of the hypochlorite ion has been depleted. In this instance, the pH will drastically drop from 7.52 to pH values closer to 2 or lower. If either of these situations occur, your hot tub will start to look grimy and you will need to add more buffer to the water (or get a new hot tub).

I know this goes without saying, but these buffers are not magical. They cannot resist tons of acidic or basic foreign objects; they can only tolerate small additions of the molecules. And this brings me to the concept of buffer capacity. Like I mentioned previously, a perfect buffer has a 1:1 ratio of weak acid and conjugate base or weak base and conjugate acid. With these ratios, the buffer will be able to resist the maximum amount of added acid or base.

Buffers will work as long as their concentration ratios remain within the 1:1 to 1:10 range. Outside of these ratios, the buffers no longer work and the pH will drastically change upon the addition of acid or base to the solutions. This is usually obvious because the color of the water changes and sometimes it starts to have a funky smell, which means it's time to clean the water in your hot tub or pool.

I like to equate buffer capacity to one's tolerance to alcohol. Picture a college freshman who has not quite figured out their limit with ethanol. At eighteen years old, they can probably take one, maybe two shots before getting too sloppy. But around shot three, we start to see their tolerance drop, and the ethanol begins to affect their basic human function. At shots four and five, the poor kid will probably black out and can no longer tolerate any more alcohol—just like how buffered water outside of its buffer capacity can no longer handle the addition of more acid or base.

In the case of a swimming pool or hot tub, the buffer remains present until it has reacted with too many foreign objects (like bacteria) that are slightly acidic or basic. It is usually recommended to check your chlorine and pH levels two to three times a week, but that seems a little excessive to me unless you're throwing a lot of house parties or if your pool is exposed to heavy rain. Otherwise, once a week is good enough for most climates.

Or you can be like my brother, who became so frustrated with his pool that he switched from the hypochlorous acid buffer to a saltwater mixture. This system uses a small electric current to break down the sodium chloride salt (NaCl) into sodium (Na^+) and chlorine gas (Cl_2). This process, called electrolysis, uses an external power source (like a battery) to force electrons to travel (unfavorably) from low to high energy.

When this happens, the sodium ions form IMFs with the water to form saltwater, while the chlorine gas dissolves to produce the previously discussed hypochlorite-hypochlorous acid. If the pH of the pool is off, the salt cell (a.k.a. the chlorine generator) will convert table salt into chlorine gas, which quickly generates bleach to keep your pool free of algae and other green gunk.

With this in mind, there is one last category of cleaners that I would like to dissect: natural cleaners. These are the natural cleaning products that claim things like "chemical-free" or "natural."

First of all, nothing is ever chemical-free. If it has an atom, then it is a chemical. And as we learned in the first chapter of this book, everything has atoms.

Second, a natural cleaner is typically a molecule that has been extracted from a plant, which doesn't necessarily mean it is better (or worse) for you than one that has been synthetically made. From a chemistry standpoint, most cleaners contain molecules that are either acids or bases. The ones that claim to use the power of lemon are really just taking advantage of the acidic properties of citric acid.

When I'm shopping for a cleaning product, I always look for items that are environmentally friendly. I don't want my cleaners to contain any phosphates or for them to release any toxic gases. I also do not want them to contain any sort of microbeads, which have become a big problem in our oceans.

But if you take all of these factors into consideration, my hope is that you will be able to pick a cleaning product that is right for you and your family, understanding that it's all a matter of acid and base chemistry that you rinse down your drains when you're done. More importantly, I hope I've scared you enough that you will never, under any circumstances, combine two household cleaners.

Now that we've got your chores (and mine!) out of the way, let's move onto something that probably brings us a lot more joy—happy hour!

11

HAPPY HOUR IS THE BEST HOUR

At the Bar

When I decided to write a book about chemistry in our everyday life, I knew I needed a chapter about happy hour. Now, as I write, bars are closed due to the COVID-19 pandemic. Still, there's no better way to end the day than hanging out with friends, telling stories, and buying cheap drinks. On sunny days in Texas, I go straight for some queso with a frozen margarita, but on date night, I tend to drink a glass of wine (or two). My husband always orders a whiskey before switching over to a refreshing beer; I can't wait for life to go back to normal again. But no matter which cocktail you're drinking, every one of them is loaded with chemistry. Let's start with the basics.

Alcohol is the generic name for a molecule that has a bond between a hydrogen atom and an oxygen atom, where the oxygen atom is directly connected to a carbon atom, like this: C–O–H. For example, methanol is an alcohol because it has the molecule formula CH_3**OH**. Ethanol is an alcohol because it has the molecular formula $CH_3\mathbf{CH_2OH}$. (The unbolded hydrogens are bonded to the carbon atoms, not the oxygen atom).

Depending on the context of your conversation, alcohol (or alc–OH–ol, if that's easier to remember) can be the nickname for several different molecules. For example, at the doctor's office an alcohol could be rubbing alcohol (isopropyl alcohol or isopropanol). In Asia, alcohol can be used as a fuel (methyl alcohol or methanol). In a margarita, alcohol is the molecule that gets you drunk (ethyl alcohol or ethanol). For that reason, I'm going to focus on ethanol in this chapter.

Oh, sweet, sweet ethanol. The way it reacts and bonds to molecules in our brain makes it a popular choice for after-work libations, first dates, and final breakups. But the process of how it is made is just as fascinating—and perhaps as surprising as why we enjoy it so much.

Historians are pretty sure that people have been making wine from grapes since 6000 BC and fermenting fruit since the Neolithic period. There is even a theory—called the drunken monkey hypothesis—that our brains have a natural attraction to ethanol because of something our ancestors did a long time ago: they ate ripe fruit (or fermented fruit) that contained ethanol, which created a natural attraction/joy response when we encountered the molecule elsewhere. In other words, in the same way that we have an inherent attraction to the smells and tastes of an apple or a banana (not because of their inherent "apple-ness" or "banana-ness" but because of the nutrients associated with them), we are also evolutionarily wired to be attracted to ethanol, once associated with those same nutrients.

We've long been experimenting with fermenting and separating those happiness-inducing ethanol molecules from the fruits and vegetables they occur in naturally. And we've got it down to a strong science, for the most part.

In general, wine is made in three steps. In the first stage, the grapes are picked from the vines, and then crushed to collect the juice. The machines that do this are actually quite delicate because they apply just enough pressure to break the skin to release the grape juice (called *must*), but not too much pressure to crush the tannins in the seeds at the center of the grape. After this, the stems are usually removed from the must because they have an unpleasant bitter taste. The remaining liquid part of the must is 12–27% sugars, 1% acids, and the rest is water. After placing the grape mixture in the proper container (with or without the skins), the best part of wine making can begin: the fermentation process.

Fermentation is an anaerobic process—or a chemical reaction that does not need oxygen as a reactant—that forms the molecule ethanol as a product. However, if oxygen is around during this process, the glucose will react to form ATP (as we learned in the fitness chapter), instead of producing ethanol.

But when oxygen is not around, yeast and sugar can react to form the alcohol. I bet you are familiar with this process if you have ever made bread from scratch before. The very first step is to activate yeast by letting it sit in sugar water for a few minutes. During this time, the yeast breaks the glucose molecules apart into smaller molecules, including carbon dioxide gas. That's why, after enough time has passed, you should be able to see a light brown bubbly concoction floating on top of the water.

When making wine, this exothermic reaction—meaning heat-releasing—converts glucose (sugar) and yeast into ethanol and carbon dioxide, as shown below:

Glucose + Yeast → Ethanol + Carbon Dioxide

The product, pure ethanol, should actually have a bitter taste and it is extremely flammable. If someone ever tries to convince you to take a flaming shot, please politely decline and get as far away from the flames as possible. That kind of recklessness can set the entire building on fire—or worse, your face—just with one slip of the hand.

But if the ethanol is not set on fire and is allowed to continue to ferment, it will turn into acetic acid (vinegar, the cleaning product). Gross, right? That is the reason why it is important for winemakers to stop the fermentation process at the right time. Otherwise, the resulting wine will have a lovely sour vinegar taste.

The type of yeast used varies, which is another reason why different wines have different tastes. Some winemakers use yeasts that naturally exist on the skins of grapes, whereas others prefer to use a starter culture of yeast.

These fermentation starters—also known as *mothers*—are basically giant bowls of good fungi that self-replicate. The single-celled microbes react with the natural sugars that exist in the grape juice to release carbon dioxide gas. During this process, our beloved ethanol is formed as a byproduct.

These yeast starters are typically kept in cold environments and can be passed down from generation to generation. There are stories of old Italian grandmothers sneaking their delicious-tasting starters onto transatlantic ships to make sure they were passed on to their grandchildren. These yeast starters were usually used for bread, but the science is the same for alcohol.

Red wines ferment (in other words, the yeast reacts with the grape juice sugars, producing ethanol and carbon dioxide) for anywhere between four days and two weeks before the skins are finally removed. They continue to ferment until two to three weeks have passed (total). White wines, however, need about four to six weeks for fermentation, since they do not have any of the natural yeasts available on the grape skins. Sometimes, if the winemaker wants to add a practice called malolactic fermentation, they do it here in the second step.

Malolactic fermentation has been around since the discovery of wine, but in the 1930s, an enologist named Jean Ribéreau-Gayon accurately described the chemical reaction that occurs when malic acid (which exists naturally in most fruits, including grapes, and gives them a mildly sour taste) is converted into lactic acid. This decreases the taste of tartness in the wine. From what I can tell, each winemaker has a very strong opinion on whether malolactic fermentation is a good or bad thing. Some encourage it by introducing *Leuconostoc oenos* bacteria into their wine, while other winemakers do everything possible to ensure it does not happen.

Ever since we invented wine, we've been playing with the process, the colors, and the tastes. Historians believe that the first wines were all red, until the Egyptians found a color mutation—and process—that produced white wines. Red wines, of course, are only light red to start, and get their deep color (and unique flavors) from sitting on the skins during the fermentation process. White wines only soak in the skins and seeds of the grapes for a few hours, and then the juice is removed.

There are also a few neat subcategories that mix the two techniques together. For example, pink wines—called rosé—come from red wine grapes that are made like white wines. The maker does not let the liquid sit on the skins very long, which is why the resulting wine has a nice pink color. Orange wines, on the other hand, are white wine grapes that are made like red wines. This wine mixture also sits on the skins during fermentation, allowing for the wine to turn into a cool orange color. The longer the juice sits on the skins, the darker the color will become.

In California, most white wines (other than chardonnay) are fermented in stainless steel, which has no interaction with the liquid inside. However, red wines (and chardonnays) are often put into barrels. The type of barrel (American oak, French oak, even barrels used for bourbon) is what deepens the flavor.

In this last step, the wine matures and develops its intense layers of flavors. This process greatly varies for each type of wine and by each winemaker, but in general, some sort of wine racking occurs. A barrel of wine is set on a big rack, kind of like what you would see in a warehouse. The barrel is moved occasionally but the purpose of racking is for any solid particles to separate from the rest of the (liquid) wine. These particles would slowly sink to the bottom of the barrel, allowing them to be filtered out. Each barrel of wine is filtered several times while it's racked.

This process of filtration is one of the most fundamental techniques used by any synthetic chemist. In the lab, we are constantly purifying our products by dissolving them in a liquid, and then filtering out any remaining solid particles. When winemakers rack wine, they are doing the exact same thing by filtering the wine over and over again, but on an extremely large scale. During this process, they are attempting to remove any remaining grape fragments and dead yeast cells.

At the very end of racking, the wine can be fined. This is when the winemaker adds a fining agent to the wine, like activated carbon charcoal or gelatin from fish bladders, which forms IMFs with the remaining solid particles in the wine solution. The resulting compound is too heavy to remain suspended in the liquid, therefore it sinks to the bottom of the container.

When the wine is finally ready, it is bottled and fitted with a cork. The best wines do not have a large gap between the top of the wine and the bottom of the cork. This is because the molecules in the wine can oxidize—or perform a chemical reaction with the oxygen gas in the air gap above the wine. Unfortunately, this is the worst thing that can happen to a bottle of wine because it causes the overly sweet smell that we associate with "corked" wine.

This is also the reason why wine is stored on its side, to prevent oxidization. When the bottle is on its side, the liquid keeps the cork wet, which allows for the cork to continue protecting the wine from the oxygen in the atmosphere. However, if the wine is stored in the vertical position, the cork can dry out, and then tiny oxygen molecules can sneak through the air pockets in the cork to get into the wine and ruin it. The smell of the cork can be used to check the integrity of that particular bottle of wine.

Once the wine has been opened, the same oxidation process will occur. That's why you may detect a difference in the flavor of your wine between day one and day two. After that, as long as you keep your wine corked, it should keep for another three to four days. But as we just discussed, all wines are different, therefore you may need to perform a small smell (or taste) test on questionable wines. Generally speaking, after five or seven days, uncorked wines are better for cooking.

These days, more and more wines are bottled with a screw cap instead of a cork. I asked a sommelier about this when my husband and I were on our five-year anniversary trip to the Basque Country in Spain. He explained to us that the selection of a screw cap versus a cork is directly related to the aging process of the wine. If the wine needs to age for a long period of time, the winemaker will likely use a cork to preserve the oxygen concentration in the wine. However, if the wine is going to be opened and consumed right away, like an Oregon pinot noir, then a screw cap is just fine.

Champagne and more expensive sparkling wines especially need to be corked because they have a secondary fermentation process. Here, the yeast is converted into ethanol and carbon dioxide, but unlike the first fermentation process, the carbon dioxide is now trapped inside of a closed container, rather than being released into the atmosphere like with a traditional wine. This process takes at least two months and may last for a few years. When the yeast cells inevitably die, they provide the sparkling wine with a distinctive roasted flavor. The solid particles are once again removed after the fermentation process is over, and then the sparkling wine is recorked.

Even inexpensive sparkling wines that are pumped full of carbon dioxide gas (rather than using the carbon dioxide produced during fermentation) and stored under pressure—in the exact same way that sodas are carbonated—need to be corked. A screw cap cannot withstand the amount of pressure within the bottle, which means that the sparkling wine could unexpectedly shoot out of the bottle, like Old Faithful, at any moment in time.

A permeable cork, on the other hand, allows the sparkling wines to slowly release carbon dioxide into the small gap between the wine and the cork. It still builds up a lot of pressure, which is why you hear a pop when you open a bottle.

When my husband and I traveled to Spain, we developed a slight obsession with cava, which has been dubbed the "Spanish champagne." It's become our tradition to start our date nights or celebrations with a round of cava, and our friends have started to notice. In fact, at a recent happy hour, we shared a few details of our trip with our friends—one of whom happened to brew beer as a hobby—which led to a great conversation about the differences between wine making and beer brewing. Of course, in the pursuit of tasty liquid + ethanol, the processes are very similar.

While wine uses the simple sugars in grape juice to react with yeasts in the fermentation process, beers use complex sugars (like starches from grains) as their starting materials. But starches need to be broken down into smaller molecules before they can be used for anything useful. So, what do we do?

We cook them.

But it's a little more complicated than that. The most common grain used in beer brewing is barley. The barley looks like small pale yellow seeds, and it is removed from its tall grasslike stalk by a huge combine harvester. The barley is then steeped in water for about two days at approximately 15°C (59°F). The goal is to get the barley seeds to be as plump as possible.

After that, the seeds are germinated for about four days (although some seeds need eight or nine days). During this process, a variety of different enzymes are generated, which immediately start to break down the cell walls in the seed. Some of the enzymes, however, instead focus on converting the starches in the barley into sugars, and the proteins into amino acids. This entire process can be observed macroscopically because the *malt*—the term used for germinated seeds—begins to darken. The longer the seeds germinate, the darker the malt.

Some brewers argue that kilning—the third and final step of making malt—is the most important part of the process. Here, the water molecules are removed from the seeds by adding a steady supply of warm air (around 55°C or 131°F). But depending on the desired malt, the seeds can be heated up to 180°C (356°F). So there's a huge range! Malts made from lower temperatures have higher enzyme activity and lighter colors, while malts from higher temperatures have low enzyme activity and deep, rich flavors. These dark malts can taste smoky, toasted, or even caramelized, even after being in storage for a few months.

In the next step, the malt is ground into a fine powder before the brewer activates those previously mentioned enzymes by soaking the powder in water, again. And again, the enzymes break down any available starches into sugars, resulting in a brown liquid—which is basically a fancy solution of deliciously sweet sugar water—called *wort*.

The wort is then boiled with hops, which gives the sugar liquid its deep, bitter flavors. For those of you that are unfamiliar, hops are small green flowers that look an awful lot like fluffy blackberries. This boiling process takes about ninety minutes, not only to make sure the wort has fully absorbed the unique flavors, but also to finally kill all of the active enzymes. Some people inaccurately think that this step is only to enrich the taste of the beer, but it really is to make sure that the number of sugar molecules in the beer remains constant. Otherwise the enzymes would eat them uncontrollably, which, as I'm sure you can guess, negatively affects the taste of the beer. The mixture is then cooled to about 10°C (50°F) before my favorite part starts—fermentation.

This fermentation process is very similar to the one we discussed for wine making: yeasts are used to convert the hops-flavored sugars into ethanol. The one major difference is that brewmasters refer to the process as either *top* or *bottom* fermentations; top fermentations give us ales and bottom fermentations give us lagers.

Let's talk about ales first. When ale yeasts are added to the wort at high temperatures, they form clusters within the mixture before floating to the top of the solution. One of the most popular ales—India pale ale (IPA)—has gained a lot of attention in the States over the past few years. These beers usually contain higher ethanol concentrations and are often accompanied by a bitter taste. Traditionally, pale ales like Sierra Nevada pale ales have fewer ethanol molecules, likely because of a shorter fermentation process. But both are top fermenters that marry yeast and very hot wort.

There is a special category of pale ales called SMaSH IPAs that are made with one part hops and one part malt (i.e., single hops and single malt). I learned about them the summer of 2020 because a local Austin brewery named their new SMaSH IPA after me (Kate la Química). They used an experimental hops, and they let me watch every step of the brewing process. I was in nerd heaven.

Another option is to mix lager yeasts with wort at low temperatures (a.k.a. *cooled wort*). They clump together before sinking to the bottom because these larger molecules are heavier than the ale yeast molecules; the beer cannot suspend them throughout the liquid (like in a colloid).

Why they don't call it hot fermentation versus cool fermentation, I'll never know. Anyway, here in the United States, most brewers use cool/bottom fermentation because it offers a drier, "bready" flavor. Since lagers usually have low ethanol levels, newcomers usually start with these types of beers, like Budweiser and Coors.

My favorite beer, wheat beer, can be made by using either top or bottom fermentation. American wheats tend to be hoppier than German wheats since they do not use the weizen yeast strain. For this reason, German wheats tend to be a bit fruitier and less bitter, and in my opinion, more delicious.

The last step of beer brewing is called conditioning, and it offers a lot of similarities to the wine purification process that I discussed earlier. This time, the fining agents are added to form bonds with the suspended tannins and proteins before being filtered out of the beer. This is also the part that removes any of the remaining dead yeast cells.

Unlike wine, beers do not need to be stored on their sides or corked. They just need to be stored in cold, dark places. Sunlight is strong enough to break the bonds in the aromatic molecules, which can release sulfur into the beer. This is what you are tasting in a "skunked" beer. Interestingly, brown glass can absorb some of the lower energy sunlight and protect beer from its harmful radiation. Green glass cannot, which is why most beer bottles are dark.

The final product—the beer—is typically 90% water, 2–10% carbs, and 1–6% ethanol. As you can see, the composition of beer greatly varies from beer to beer. That's why we use the terms ABV or *proof* to describe the concentration of ethanol in an alcoholic beverage. ABV is the percentage of alcohol by volume, which is a concentration term that irks me so much because it should really be *ethanol* by volume.

The term *proof* was originally used in England as a way to apply different tax rates to spirits over beers. The alcohol was poured over gunpowder and ignited. If the beverage burned at a steady rate with a specific blue flame, it was "proven" to be a quality alcohol. However, if the drink did not burst into flames, it was underproof, which means it did not contain enough molecules of ethanol. If the gunpowder caught fire too quickly, the alcohol contained too many ethanol molecules, thereby indicating that the spirit was overproof.

Nowadays, ABV is considered to be the most common way to indicate ethanol content in an adult beverage. If you see *proof* on a bottle in the United States, just divide the number in half to get the ABV. The National Institute on Alcohol Abuse and Alcoholism reports that a twelve-ounce beer is on average 5% ABV so we'll use that number. For comparison, an average five-ounce glass of wine has between 12 and 18% ABV. And sake is closer to 20%!

Why is that? Sake is essentially a half-wine, half-beer alcoholic beverage. Although its production follows more closely to that of wine, it doesn't originate from grapes or any other fruit for that matter. Instead, sake—like beer—is made from a grain, in this case, rice. During the fermentation process, a sweet *mold* is added to the rice as a way to provide the solution with enzymes that breaks the rice starches into sugars. At the same time, yeast is added to this mixture to react with the sugars to form the ever-so-important ethanol.

But unlike beer, this pure fermentation method can generate a potent liquid with an ethanol content of 20%. One of the reasons sake has a significantly higher ABV than beer is because cooked rice is continuously added to the solution during the fermentation process. The mold used to convert sugar into ethanol also happens to grow naturally on rice. Therefore, adding cooked rice is actually twice as beneficial.

Sake is considered to be the most fragile of the three alcohols I've discussed so far. This is because it does not have any of the colorful molecules found in grapes or hops to absorb incoming sunlight that would otherwise damage the molecules responsible for the light, flowery flavors. The clear and blue glass bottles used by the Japanese sake industry are also useless against sunlight. This is particularly ironic because sake is even more vulnerable than wine or beer.

For this reason, it is usually recommended to drink sake quickly, especially after the bottle has been opened. However, bearing in mind that it has a 20% ABV, perhaps it's better to share with a large group of friends.

But don't offer any to me, please. The ethanol content is so high that one whiff of it immediately makes me think of the lab.

I have the same exact reaction to vodka, which is basically just watered-down ethanol. One of the most potent vodkas, Spirytus, comes from Poland. It is 96% ABV or 192 proof—or 96% ethanol. You could not pay me to try that vodka.

Most vodkas are usually around 40% ABV, or 80 proof, because they all go through a similar purification process. Like wine, sake, and beer, liquors all start with fermentation, but they also go through an additional step called distillation. This process is an extremely crucial part of making spirits and liquors because the starting materials often include a dangerous molecule called methanol.

The molecular formulas of methanol and ethanol are very similar, but the way they function in our bodies are very different. When we drink ethanol (CH_3CH_2OH), we get drunk. But when we drink methanol (CH_3OH), we go blind. The methanol turns into formic acid, which negatively interacts with the optic nerve to cause blindness. Have you ever heard the expression *blind drunk*? Now you know where that came from.

Thanks to institutions like the FDA, we no longer have to worry about the methanol in our cocktails. But during Prohibition, many new chemists—bootleggers—started making ethanol in their kitchen. The problem was the majority of these people did not have a strong background in science and accidentally wound up making methanol, creating a blindness epidemic. For that reason, it's good to think twice when offered a buddy's backyard moonshine.

Too much methanol will actually kill you, which is why vodkas (and whiskeys, scotches, tequilas, and rums) are always distilled. To purify vodka, a high-alcohol liquid made from fermented grains (like potatoes or sorghum), the spent yeast is removed so just the ethanol/methanol mixture remains. This solution is then slowly heated at different temperatures over a period of time.

In the beginning, the mixture will spend some time around 65°C (149°F) so that all the methanol boils and eventually vaporizes (turns into a gas). This allows the maker to remove the gas and therefore, the methanol too.

After that, the temperature is then increased to around 78°C (173°F) to collect the ethanol from the sample. Once the ethanol vapor leaves the alcohol solution, it travels up a fancy glass tube called a condenser. Here, the gaseous ethanol condenses back to liquid ethanol and drips down into a new container.

This process is repeated two more times for triply distilled vodka. In goes the ethanol mixture (with some methanol contaminants that sneaked in), on goes the heat, and then just like Superman, the gaseous methanol molecules go up, up, and away (once they reach their boiling point).

Not only does the distillation process purify the alcoholic sample, but it also inevitably increases the concentration of the ethanol in the sample. But since the distillation process makes the vodka extremely pure at 40% ABV, what is the remaining 60% of the solution?

Water.

That means you are actually staying hydrated while drinking. Just kidding. You are definitely *de*hydrating yourself. That's why you get a hangover if you drink too much.

That said, the ethanol we drink in vodka is always combined with water. These two liquids are miscible—they mix well together—because they form hydrogen bonds between the oxygen atoms on water and the hydrogen atoms in ethanol (and vice versa). Liquids that are immiscible form two different layers when mixed together, like oil and water. Luckily, most mixers are miscible with ethanol and water.

But if you've ever had a B-52 shot, you know that it's possible to stack three different alcoholic liqueurs on top of each other (coffee liqueur at the bottom, then usually Bailey's Irish Cream, followed by Grand Marnier on top). This is just another quick example of how densities work.

I am usually open to trying most liquors and spirits, but there is one that I just cannot bring myself to taste: absinthe. My manager and I recently went to an oyster bar in Brooklyn that was serving absinthe as part of their happy hour special.

The waitress was pushing us to try it, and I was politely trying to tell her "hell no."

Absinthe is a spirit that has not been properly defined by most countries, therefore it is able to dance around traditional alcohol regulations. Unlike other spirits like brandy and gin, the name *absinthe* can be applied to a variety of different beverages since all countries (except Switzerland) have yet to legally define the spirit. That is another reason why you can find absinthe bottles made of 45% ethanol and up to 70% ethanol.

This misunderstood libation comes from three main plants, including wormwood, fennel, and anise. The wormwood plant is actually quite bitter, which is one of the reasons why absinthe is typically poured over a sugar cube. Some places even set the sugar on fire (really, the alcohol vapors from the absinthe), which adds a toasted flavor to the absinthe drink—and another reason why I avoid it.

Before I go into how it is produced, I want to address a rumor about absinthe. Absinthe is *not* a hallucinogen or psychoactive drug. As far as we can tell, all of the stories about absinthe causing people to experience hallucinations or commit violent crimes were just that: stories. From a scientific point of view, there is no reason why absinthe would cause the human body to freak out like that.

However, for nearly seventy-five years, a molecule named thujone was blamed for causing people to think and act in erratic ways. It was believed to be toxic to the human nervous system, and it was known to cause convulsions. This molecule is present in wormwood oil and was reported to be the reason why people were having negative responses to absinthe. However, it was later discovered that Dr. Valentin Magnan, the scientist who made these connections between thujone and absinthe, was actually a teetotaler of sorts and opposed alcohol consumption in France.

As one might expect, recent discoveries have proven that they could barely detect any thujone molecules present in the absinthe samples preserved from the twentieth century. *How* Dr. Magnan was able to convince the entire world about the "dangers" of absinthe, we'll never know. But he started an incredible rumor that absinthe has never been able to kick.

Since Switzerland is the only country currently to have a proper definition for absinthe, let's look at their production process a little more closely. In Switzerland, absinthe is produced by soaking the anise, wormwood, and fennel in a solution of 96% ethanol. This process is called maceration, where the botanicals absorb the ethanol through a semipermeable membrane. At the same time, the ethanol picks up the light, flowery flavors of the herbs.

Just like with vodka, the absinthe is distilled to purify the liquid. This immediately drops the ABV down to 70% or so, once the methanol is removed. Some makers dilute it down further just by adding more water to the solution. After this, the absinthe usually goes through maceration again, but this time with hyssop, melissa, and more wormwood. This part is very similar to the way we make tea—except here, the ideal final solution contains a lot of chlorophyll. And that one chemical is how absinthe gets its notorious green color.

It is true that a few drinks of absinthe, like all high-proof beverages, can really mess a body up. That high of an ABV can cause a series of chemical reactions that greatly affect how the human body can function. This is because alcohol is metabolized differently than food. Initially, the ethanol is absorbed into the veins from the small intestine before being transported directly to the liver. (Hence why we measure blood alcohol levels for inebriated driving.)

Once ethanol has reached the liver, an enzyme called alcohol dehydrogenase (ADH) breaks two covalent bonds in the ethanol molecule (CH_3CH_2OH). This process, called the partial oxidation of ethanol, forms acetaldehyde (CH_3CHO). You'll learn to hate acetaldehyde in a bit, but for now, just know that it quickly converts into the acetate (CH_3COO^-) ion.

A different enzyme called aldehyde dehydrogenase comes in and then breaks the acetate ion down into carbon dioxide and water before we exhale the carbon dioxide out. That all sounds rather innocuous, doesn't it? If you didn't know any better, it would sound an awful lot like what happens when our stomachs digest food.

So, what's the science behind getting drunk? What is it about ethanol (CH_3CH_2OH) that makes me want to put on my cowboy boots and do the Texas two-step after a few drinks at happy hour? And why is it that drinking on an empty stomach seems to make that happen a little faster?

For starters, the ethanol reaches your brain after about five minutes of ingesting it, and five minutes after *that*, you might start feeling it. At that moment, your brain has had enough time to interact with the ethanol to start releasing dopamine. In your brain, dopamine acts as a neurotransmitter, or a molecule that carries signals from one nerve receptor to another. When we drink alcohol, our brain reacts by releasing dopamine molecules.

Dopamine carries a reputation for being a superhero molecule that immediately makes a person feel "happy." However, as we discussed in the fitness chapter, research has shown that dopamine molecules don't actually provide the person with motivation toward (or against) doing something. In this context, dopamine signals the perceived positive aspects of doing something—like drinking alcohol—which is why people automatically associate dopamine with pleasure.

Ethanol also interacts with our sodium, calcium, and potassium channels and messes with another neurotransmitter in our brains, GABA. This one, however, inhibits brain activity, and we can usually watch the direct effects of this on a drunk person. GABA is responsible for the clumsiness these people exhibit, and the stereotypical slurred speech. Not to be confused with the bad dancing usually associated with ABBA.

In addition to GABA, ethanol disrupts the way another neurotransmitter functions. Glutamate is an important molecule that plays a key role in the way our brain forms memories and ultimately learns new things. When the movement of glutamate is inhibited, the inebriated person begins to have a hard time trying to learn new things—or form new memories. I swear some people have some kind of special glutamate blockers because they cannot remember a thing after one or two drinks. It is seriously remarkable. This is also what can happen after a night of heavy drinking.

All three of these factors combine to give us a drunk person in a state of euphoria who trips on their own feet and will not remember any of it in the morning. And of course, there are degrees of severity depending on how much you drink.

We refer to the amount of ethanol in your body as the blood alcohol content (BAC), which is a number that indicates the percentage of ethanol that has actually been absorbed into the bloodstream. The legal limit is usually 0.08 for driving a vehicle. The reason being that the human body begins to truly feel the effects of the GABA around 0.09 up to 0.18.

At a BAC of 0.19 and higher, the body usually begins to experience confusion. This is more so due to the glutamate blocking your ability to form new memories. These BAC levels are also associated with a lack of coordination, which are definitely attributed to the presence of the GABA neurotransmitter. If you've ever "blacked out" before, that's because you likely had a BAC greater than 0.19, which is bananas.

When the BAC gets above 0.25, the person can begin to show symptoms of alcohol poisoning. This is where danger really sets in because a person in this state could easily choke on their vomit and asphyxiate.

If you somehow drink yourself to a BAC of 0.35, you run the risk of falling into a coma. You also begin to have issues with basic circulation and respiration. A BAC of 0.45 is an almost guaranteed death. The brain can no longer function properly enough to instruct the body how to do basic things, like take a breath. That being said, Sweden claims to have a record of one of their citizens with a BAC of 0.545!

Think about what it would mean to have a BAC of 0.545. If we assume that the average human body has around five liters of blood, a BAC of 0.545 indicates that their blood alone would contain an entire ounce of pure ethanol. That's equivalent to a 2.5 ounce shot of vodka—in your bloodstream! When judges order a warrant to collect a blood sample from a drunk driver, it's so scientists can quantify the amount of ethanol present in his or her blood.

But if we actually need a sample of blood to determine your BAC, how can policemen detect the number of ethanol molecules in your body on the roads?

They can't.

That's why they have to use a breathalyzer. The majority of these machines contain some fascinating chemistry that convert the ethanol (CH_3CH_2OH) vapor in your exhale into acetic acid (CH_3COOH) in a very short period of time. This is the exact same chemical reaction that we discussed earlier when people let their home brew ferment for too long—the ethanol converts into acetic acid (vinegar).

Because acetic acid does not naturally occur in our atmosphere, any acetic acid detected by the breathalyzer can be attributed to the amount of ethanol in your body. Therefore, the breathalyzer can do a quick calculation to spit out a relatively accurate number for your BAC.

The small breathalyzers that many police officers carry in their cars are not very accurate, thereby they are often not usable in court. However, they are good enough to justify your arrest. From there, the police officer will take you down to the hospital to collect a blood sample.

An average Joe can purchase a breathalyzer to have in his or her car, therefore anyone can test themselves after a couple drinks at the bar. If you blow over 0.08, you need to call an Uber or a Lyft (or a friend) to drive you home. In this day and age, it's too easy to find another ride home instead of putting your life—and other people's lives—at risk. GABA is not our friend after a few margaritas.

All of this disruption of neurotransmitters, while fun in the moment, can have detrimental effects on how you feel the next morning. The hangover, besides being one of my favorite movies, has got to be the worst part about having a few drinks with friends. Not only is your body now dehydrated from processing all of that ethanol, but the enzymes in your liver have broken down the ethanol and created a buildup of that molecule I mentioned earlier, acetaldehyde.

Acetaldehyde is essentially an intermediate in the synthesis of other chemicals—the primary one being the acetate ion, which is eventually broken down into carbon dioxide and water—but the process takes some time. The toxic acetaldehyde molecule remains in your body for hours (after all of the ethanol molecules have been removed) before it is further metabolized.

This is especially true for certain types of people that carry a mutation to their alcohol dehydrogenase gene. For this cohort, the ethanol is converted into acetaldehyde extremely quickly, but then their body struggles to continue the transformation to the acetate ion. When this happens, the body becomes all red and splotchy due to the alcohol flush reaction. Red patches can show up right away or later in the evening and are usually a good indication that a hangover is coming.

One of the worst parts about a hangover is the electrolyte loss associated with the chemical reaction of removing alcohol from the human body. Electrolytes, which you may recognize as minerals from the breakfast chapter, can be separated into two categories: cations (positively charged ions) and anions (negatively charged ions). The primary cations in our bodies are calcium, magnesium, potassium, and sodium, and our anions are chloride, hydrogen carbonate, and hydrogen phosphate. In general, we maintain a 1:1 ratio of cation:anion electrolytes within our bodies.

On non–happy hour days, our kidneys work to maintain the proper concentrations of each of our electrolytes. This is extremely important because these minerals regulate the pH of our blood and are responsible for the contractions of our muscles (like our heart). But on happy hour days, our kidneys will jump into overdrive. Why is that?

Ethanol is a diuretic, which means it makes you have to pee, a lot. Every time you go to the bathroom, you push necessary electrolytes out of your body. This is why a Gatorade the morning after drinking can sometimes sound better than chocolate chip pancakes. It's a quick and inexpensive way to replenish the potassium and magnesium (among other ions) needed to fight off the fatigue and aches attributed to drinking too much the night before.

The best way to beat a hangover is to be a responsible adult. You already know to make sure your body is properly hydrated before you go out for a few drinks. But you should also have dinner with your first glass of wine and make sure to eat food that takes a while to digest. The ethanol actually has to pass through the food before it can be absorbed into your veins. Therefore, the complex carbs that we discussed in previous chapters—like potatoes and corn—can act like a physical barrier in this context, slowing down the absorption of ethanol. So make sure you eat!

Now, what have we learned? All alcohol is the result of fermentation, and all spirits are a result of fermentation plus distillation. Winemakers have strong feelings about malic acid, and beer brewers love their wort. And for most of us, it's perfectly fine to have a few drinks at a happy hour with friends, eat as much queso as possible, and then maybe even try the Texas two-step. Just don't forget that you've got to filter all those ethanol byproducts out sooner or later, when Dr. Jekyll turns into Acetaldehyde.

12

SUNSET & CHILL

In the Bedroom

Whoever said "nighttime is the right time" must have been a chemist. In fact, some of the most exciting chemistry of the day happens in the evening. The light shows we all enjoy at sunset, the pleasure we feel after sex, that candle you used to set the mood—there is no denying that nighttime has a special energy, and it's all thanks to the interaction of atoms and molecules.

Let's get started with one of the most exquisite wonders of the natural world: a sunset. When the sun goes down, the earth cools as heat dissipates from the ground, no longer fueled and energized by the sun. The beginning of that transition is spectacular: the slow fade from light to dark, the beauty of the sunset. And if you happen to wind down a little early, especially in the middle of the summer when the days are the longest of the year, you may be fortunate enough to catch one of the most stunning phenomena nature has to offer: crepuscular rays. These "God rays" are produced when light reflects off dirt particles—or molecules—that are floating in the air, and have the effect of seeming as if a celestial being is splitting the clouds to shine a spotlight down onto Earth.

I remember first noticing crepuscular rays when I was a kid at my family's cottage in Michigan. We have a small house on a little lake that has two or three feet of sandy beach all around it, where my dad has hung an old hammock between two massive oak and maple trees. If you lay in that hammock at any hour, you could feel a gentle breeze from the lake and hear the waves crash into the beach. It was so peaceful and relaxing, I cannot tell you how many times I found one of my family members asleep there. It was like a tiny piece of heaven. And when crepuscular rays shined through the clouds, it actually looked like heaven too.

These sunbeams, also affectionately referred to as "Buddha rays" or "Jacob's Ladder," are distinctive patterns of sunshine that alternate between light and dark streams of light. Each pattern is unique because their shape depends on the positioning of the clouds and the time of day (as well as the positioning of the sun). They commonly appear during twilight hours because that is when the sun is below the horizon, just after the sun has set or is just about to rise. At this angle, light can easily be scattered to achieve beautiful sunrises and sunsets. But how does that work?

We already know that the sun is shining beams of light onto the Earth in the form of electromagnetic radiation (ultraviolet, visible, and infrared). If a molecule interferes with either the electric or magnetic waves, the atoms either block or bend the light. But at twilight, the dark shadows cast by clouds and mountains now run parallel to the glowing sunbeams—just like your shadow gets longer the later it is in the day. In fact, if you were to look from the top down at these parallel patterns (like from an astronaut's perspective), the dark shadows would be sure to catch your eye before the sunlight. But on Earth, we can only see the way the light pushes its way through the clouds.

For this reason, astronauts at the International Space Station (ISS) have taken pictures to show us ground-walkers how the light curls around the clouds to cast parallel shadows onto the Earth. The pictures are really neat because the clouds almost look like they have a shadow dust trail, just like meteoroids.

As I mentioned before, these beautiful sunbeams are only formed when light from the sun is scattered by the small molecules (like nitrogen, oxygen, and carbon dioxide) and pollutants (like dog hair, dust, and car exhaust) that are commonly found in our atmosphere—in the air we breathe. As one might expect, cities with larger populations tend to have significantly more particles in the air than less population-dense areas, which is why city folk (like me) often think that rural air smells so much cleaner. There are simply fewer dust molecules in the air that my lungs need to filter away from the oxygen.

When the sunbeams shine through the clouds at a low angle in the troposphere (the layer closest to the Earth), the trajectory of the light intersects with the matter in the air to cause something called an optical phenomenon. Luckily for us, the science is *very* similar to the chemistry we discussed in the beach chapter. Physicists just use the phrase optical phenomenon to describe chemical interactions in the atmosphere. Silly physicists.

These phenomena, like a rainbow or a mirage or a crepuscular ray, are grouped together into a category of interactions of light and matter that humans can physically see with our naked eyes. Reflection and refraction are also in this category and are responsible for the amazing colors of sunrises or sunsets; we'll talk about those in a little bit.

In general, optical phenomena happen when the molecules in the atmosphere are hit with lower energy light, like infrared radiation (IR), which is the weakest type of light energy that the Earth receives from the sun. Infrared radiation may be weak, but we get a LOT of it: seven times more IR than UV light on a daily basis. Fortunately, it is not strong enough to cause skin cancer (like UV radiation).

I already shared with you how infrared radiation was first discovered back in 1800 by William Herschel, but now I think you are ready to take a deeper look at this chemistry (it's the last chapter of the book after all). Remember, this discovery was before chemists and scientists had deduced that light had a particulate form as well as a wave form. But here, it's the wave form—a.k.a. the energy signature—that is important. Infrared radiation is big enough not to be dangerous to humans, but it's still quite small: it has wavelengths from 740 nm to 1 mm, which is about the size of a point of a needle. Since this type of energy is invisible to the human eye, Herschel had to figure out a way to use a thermometer and a prism to detect heat energy.

Although we cannot see this light without night vision goggles, we can definitely feel it in the form of heat. As I mentioned in our previous discussion about baking, IR energy is really just thermal energy, which is why we use it in our ovens.

Also, as with baking, when molecules interact with IR, they absorb the energy and begin to vibrate. For example, if someone sprays you with a water hose, you might jump around a little bit in reaction to the water. This is exactly what happens with certain molecules interacting with IR. The molecules absorb the infrared radiation (in our metaphor, water), and then vibrate (jump around) because of the extra energy that has just been added to their system.

Molecules like carbon dioxide and methane all react this same way when they interact with infrared radiation in our atmosphere: they vibrate, and then something cool happens.

Unlike what we discussed earlier with UV light, once these atmospheric molecules interact with lower energy light like IR, they can reemit the energy back into the atmosphere in a *different* direction. This helps our planet maintain a temperature that is safe for human survival.

To explain this a little further, let's go back to our water/hose example. If someone hits you unexpectedly with a stream of water, you are likely to jump back immediately and shake around a little bit. During this process, you may turn the orientation of your body 10° or 20°, maybe even 180°—just like the molecules do when they absorb infrared radiation.

This new energy (water from hose) causes the molecules to vibrate (jump around), which gives them a slightly different orientation in space. When they are done reacting to the surprise of the IR light (water), they will emit the energy back in whatever direction they are now facing.

In the instance of dust particles, the molecules can only hold this energy for a short period of time before they have to reemit it back into the Earth's atmosphere. This reemission of light occurs at a completely different trajectory, and when this happens under the perfect conditions (like at twilight), we get gorgeous sunbeams that shine a spotlight down onto one section of Earth.

Crepuscular rays are generally only present in the form of white light. This light appears colorless to us because it contains a perfect blend of all colors in the visible spectrum (if that feels counterintuitive, you can prove this to yourself by taking a prism to sunlight and watching it split into a rainbow).

We use the phrase white light as the generic term for the region of the electromagnetic spectrum that has a range of wavelengths of energy from 380 to 740 nm. This region, called the visible region, is aptly named because it is the exact range of light that our eyes can see. For example, any colorful substance contains a section of a molecule that our eyes interpret as a specific color. This special part of the molecule is called the chromophore, and it can absorb lots of different wavelengths of light except one.

You may be familiar with this chemistry if you frequently have to visit the eye doctor (like I do). Turns out, we have a molecule in our eyes that also contains a chromophore. It is called retinal (a form of vitamin A) and it is a wonderful molecule that helps us see. When light hits the retinal in your eye, the retinal molecules react by moving from the cis to the trans conformation, which ultimately straightens out the molecules. (Remember that cis is the configuration that is formed when the atoms are on the same side of the bond, while trans is formed when the atoms are on the opposite sides of the bond). This movement applies a pressure to the opsin protein in your retina, which begins a process that eventually allows your brain to interpret the images of the things around you.

Each of these retinal-protein interactions respond to different wavelengths of light. If the light has a wavelength between 625 and 740 nm, then our eyeballs interpret it as the color red. Shorter wavelengths of 590–625 nm are orange, followed by yellow (565 nm), then green (500 nm), cyan (485 nm), blue (450 nm), and violet (380 nm). Violet has the highest energy in the visible spectrum, therefore it has relatively small wavelengths.

However, if white light contains all of the colors of visible light, how come our sunsets are mostly combinations of pinks and reds, and occasionally some shades of orange?

In order to answer this question, we have to remember that wavelength is inversely proportional to energy. This means that light beams with long (big) wavelengths are much weaker than light with short (small) wavelengths. Blue light waves are much stronger/shorter than red light waves, therefore they will scatter much more efficiently than red light.

This concept is easiest to understand with the following analogy. Imagine the light beams, with all their different wavelengths, are bouncy balls. And we're going to throw those bouncy balls down a super old brick road. To start with, let's assume you use a lot of force to throw one blue ball down at the uneven surface (to represent blue light). As expected, it will bounce back up, but at a completely different angle and with quite a bit of speed.

Now, let's gently drop one red ball on the jagged brick road (to represent red light). You're going to use a lot less force this time, because red light is weaker. Just like the blue ball, the red ball will change trajectory, but at a much slower speed.

But now, let's consider what would happen if we dumped hundreds of red and blue balls onto the brick road all at the same time. In this instance, the blue balls would have much more energy than the corresponding red balls, and they would overpower the weaker red trajectories, essentially knocking them away and bouncing over and over again. To the human eye, the majority of what we would be able to see are blue balls flying everywhere, with a few hints of red here and there. This is what happens during the daytime—and why the sky is blue.

However, during a sunset, the sun is low on the horizon, and the light beams have a much longer distance to travel to get to you. The sunlight interacts with significantly more molecules, and (surprisingly) this results in beautiful orange and red colors in the sky.

Here's what's happening: Do you remember how oxygen and ozone absorbed UVB and UVC light by breaking their bonds? But UVA light could then sneak by because it was too weak to break the covalent bonds in the oxygen-containing molecules? The same thing happens with visible light.

The purple and blue waves are high enough in energy that they are scattered by the molecules in the atmosphere, like nitrogen and oxygen. These waves are absorbed by the molecules and the energy is then reemitted back toward the sun (away from us on Earth's surface). The orange and red waves are too weak to be absorbed; therefore, reds, pinks, and oranges wiggle by the molecules in the air, giving us stunning sunsets.

In cities where there is a high concentration of pollutants in the air, the blue light is even more aggressively scattered away from the Earth's surface, leaving only the light with longer wavelengths in the atmosphere (reds). For this reason, horrendous wildfires—with their airborne detritus—are usually accompanied by some truly stunning sunsets. In some extreme cases, like after the spring 2020 wildfires in Australia, the sky turns completely red, creating an eerie, apocalyptic feeling like a scene from *Mad Max*.

But even a standard Tuesday sunset, with a day of clear sky, can result in epic swirls of color. When the conditions are right and the light has scattered to produce gorgeous red and orange colors in the sky, the sunset can provide a picturesque backdrop for the start of a great date night.

I hope you know where I'm going with this. Because when it comes to evening intimacy, there is a lot of science that comes with setting the perfect mood. First, let's talk aphrodisiacs. Candles, chocolate, and oysters—everyone has their preference. But is there any real chemistry behind the "chemistry"? Or are aphrodisiacs only figments of our fertile imaginations?

The answer might surprise you. First, a clarification: these so-called sexual triggers earned the name aphrodisiac after the goddess of love, Aphrodite, and they should never be confused with the things on the other end of the spectrum, like garlic and body odor, that can dampen your sexual desires. These smells are aptly called anaphrodisiacs.

Aphrodisiacs vary in different parts of the world. They can range from pumpkin seeds (Mexico) to cobra blood (Thailand) to a crab smoothie (Colombia). One of the common ones in the United States is a scented candle.

There's a small candle shop in my hometown in Michigan, the Kalamazoo Candle Company, that has amazing fragrances like Moroccan rose, arboretum, and my favorite, lemongrass, all of which can be used to kick-start the fun part of the evening. And for any sapiosexuals out there, the science of candles can really get you going.

When candles are made, a cord of cellulose (cotton) is coated in hot paraffin (a wax made from hydrocarbons) before draping it over a cold surface. The instant change in temperature allows the liquid paraffin to turn into a solid, forming a protective coating around the cellulose. This is the candlewick.

To make the candle itself, a number of different methods can be used. One of the most common practices is called candle pressing, where sprinklers shoot hot (liquid) paraffin wax vertically into the air of a refrigerated chamber (lower than 25°C or 77°F). Once in the air, the hot wax instantly cools into cold (solid) wax droplets that rain back down onto a large tray. This step truly looks like a mini paraffin wax snowstorm right in the middle of a giant industrial machine. It is so pretty.

These wax flakes are then pressed together in a mold, during which the nonpolar molecules form dispersion forces with the neighboring molecules to give us the classic tubular shape of a candle.

For a scented candle, all manufacturers have to do is add some fragrant molecules like 4-hydroxy-3-methoxybenzaldehyde (vanilla) to the liquid paraffin wax before shooting it into the cold air. However, since the wax is composed of nonpolar molecules, we can only dissolve other nonpolar molecules into the paraffin mixture. If we try to add lots of fragrant polar molecules to the nonpolar mixture, they will separate before we can shoot them into the air to form the solid wax droplets.

Luckily for us, the molecules we use in candles are extremely potent, therefore we only need a few drops to give a candle some fragrance. With enough stirring, you can encourage a few drops of even the most polar molecules to disperse within the nonpolar solution. Once the wick has been added to the candle, we can use the combustion reaction to release any dissolved aromatic molecules, and let the aphrodisiac do its thing.

However, if I'm being completely honest with you, the science behind aphrodisiacs and anaphrodisiacs is a little shaky because, as it turns out, there is no solid evidence that the molecules in oysters, pomegranates, or chocolate have a direct effect on our sexual behaviors—at least not one that we've discovered. Instead, it's believed that the aphrodisiac action in these foods is merely the result of the placebo effect, which occurs when people lean in to a preexisting belief.

BUT, in actuality, there is one aphrodisiac that has been proven to affect the chemical—and sexual—reactions in your brain: ethanol.

Beer, wine, and liquor are indeed classified as aphrodisiacs because of the way the molecule ethanol alters your brain chemistry. If you are in a safe place with someone you trust, ethanol makes it easier to let your guard down and open yourself up to new, adventurous activities...maybe in the bedroom.

Of course, context still matters, and alcohol does not *always* put you in the mood. For example, when I have a couple drinks with my colleagues, I don't feel the urge to get frisky, whereas after a few cocktails with my husband... Well, different story.

When it comes to aphrodisiacs, your environment (and your company) matters. Scented candles do not necessarily get you in the mood, but if your partner gives you the look and initiates the no-pants dance, then your brain is able to deduce the pertinent information. But why is that?

It has to do with our hormones.

I've already talked about hormones in a few previous chapters, but their importance can't be overstated. Hormones are "activating" molecules that are produced by a number of different glands in our bodies. Just look at our previous examples: TSH (the hormone that affects the thyroid), epinephrine (the hormone responsible for adrenaline rushes), and cortisol (one of the two hormones behind stress). In fact, our bodies produce over fifty different types of hormones. Most are steroids (like cortisol) or peptides (like TSH), but a small number of them are derived from amino acids, like epinephrine.

As we've discussed in previous chapters, hormones have a variety of different physical properties. Some are more soluble in water (blood), while others are more soluble in fat (lipids). Hormones can contribute to our sleep patterns, our moods, and a few of them help to determine if you are *in* the mood. Take testosterone, for instance.

The story of how testosterone was discovered isn't sexy at all. In 1849, a German zoologist named Arnold Adolph Berthold was observing some of his chickens. He noticed that the castrated roosters behaved differently from the regular roosters. So, he did what any scientist would do, and decided to run an experiment on six male chickens. He removed the testes from four of them and kept the other two intact. Then, he simply watched the chickens grow up.

He noticed that the two regular roosters developed normally with typical male chicken behaviors. This included a strong crow, sexual behaviors, and appropriately sized wattles and combs. In humans, wattles and combs are equivalent to when young men develop their Adam's apple during puberty. Interestingly, the other four chickens in Berthold's experiment (the ones missing testes) *never* developed their cool wattles and combs.

That is, until he decided to do something a little...nutty. He selected two of his castrated roosters and implanted testes into their abdomens. Over time, both of the remasculated chickens developed the characteristics of a typical rooster. Berthold was ecstatic about this result because it meant that the testes were secreting molecules from the organ into the bloodstream, which was triggering the rooster to initiate puberty. He confirmed this conclusion during the autopsies of these roosters, during which he observed that new blood vessels had formed around the implanted testes.

He didn't know it at the time, but he had just discovered the hormone testosterone (the primary male sex hormone). This big molecule is responsible for secondary sexual characteristics in males, so in humans, it triggers Adam's apples, facial hair, denser muscles and bones, and deepened voices. Eventually, scientists figured out that it can also play a role in preventing osteoporosis.

Then in 1902, two English physiologists, William Bayliss and Ernest Starling, took the research one step further when they realized that hormones like testosterone operated like chemical messengers; a kind of postal service for the body, carrying chemical "messages" from one place to another.

There are lots of things that can trigger hormonal release. Many of them are environmental; in other words, various glands are wired to release "messages" via hormones, when certain conditions or actions are met. My favorite of these is oxytocin, the love hormone.

Oxytocin is a large peptide with a molecular weight of 1007 g/mol. It is made from a combination of eight amino acids, in a very specific order. Cysteine is the only amino acid that is duplicated in the chain, making the peptide a nonapeptide—or a chain of nine amino acids. Oxytocin is produced and released by the pituitary gland, which is located just behind the bridge of your nose.

There are a number of external triggers that can cue your body to release oxytocin, like when your partner wraps you in tightly for a hug or when they make your baby laugh. Your body is hardwired during these positive human interactions to flood your brain with the hormone oxytocin, which can make you feel like your heart is bursting with love. It doesn't matter if you're interacting with a romantic partner or caring for your children; the hormone doesn't discriminate. It just responds to (and arguably creates) those feelings of love. (And just to clarify, this is a different species than the bliss molecule, anandamide, the molecule that blocks pain.)

The love molecule is a very important hormone because it also manages the function of our sex organs on which babies and sexual intercourse both depend. We know that it is secreted from the pituitary gland into the bloodstream when a woman's uterus contracts during labor or when her nipples are stimulated during breastfeeding. The most oxytocin a woman will ever experience in her lifetime is during labor when her oxytocin levels are 300 times higher than normal.

Due to its effect on the uterus, medicines containing high concentrations of oxytocin, like Pitocin or Syntocinon, can be given to women to induce labor. This effect was realized in 1906 when British pharmacologist Sir Henry Dale isolated the hormone from the human pituitary gland and injected it into a pregnant cat—who immediately gave birth. He later named this molecule oxytocin for "quick birth," and it has been utilized in labor and delivery rooms ever since.

It wasn't until 1953 when an American biochemist named Vincent du Vigneaud made a groundbreaking discovery when he figured out the structure and arrangement of the amino acids in oxytocin. He proved his discovery by synthesizing the hormone in his laboratory, which had never been done before. It was such an impressive feat that he won the Nobel Prize in Chemistry in 1955.

And then about fifteen years ago, the Swedish doctor Kerstin Uvnäs Moberg published a book called *The Oxytocin Factor.* In it, she suggests that oxytocin has the opposite effect on the human body than the fight-or-flight response. Instead of making us weary and cautious of strangers, oxytocin makes us feel safe and trusting. She supports these claims with research that has been performed on animals, like rats and voles (which look like cute hamsters). For example, voles can be manipulated into selecting certain vole partners if they are injected with oxytocin when in the vicinity of a target mate.

In humans, most evidence supports the claim that oxytocin has a large effect on how we bond with each other (and even with animals). For example, when we pet dogs, scientists can detect a major spike in our oxytocin levels. This is especially true for baby animals, such as when an adorable puppy climbs into your lap for cuddles. Not surprisingly, we can detect the same spike in oxytocin when a new mom holds her baby. Chemically, there is so much love pouring out of the mother, it results in extraordinarily high levels of oxytocin in the woman's body. This love molecule earned its name for a reason.

Researchers have also noticed that the oxytocin levels spike when adults are affectionate with each other. For women, the concentration of oxytocin molecules starts to increase during foreplay. There is evidence that we generally feel more bonded to our partners when sexual encounters are prolonged, even before full-on penetration. From a chemistry standpoint, this occurs because more oxytocin molecules are pumping through our bodies.

Women get a second spike of oxytocin right after their orgasm. Physiologically speaking, this happens so that we can form a strong bond with our partner, just in case we get pregnant. The female body performs this action instinctively and unconsciously, to help solidify the bond between the two people.

Men, on the other hand, do not get to enjoy the second oxytocin spike. Instead, they have a general increase in oxytocin during all types of sexual arousal, which plateaus after orgasm. Researchers believe that the lack of secondary oxytocin boost is because there is no physiological reason for men to form a strong bond with their partner, since they cannot get pregnant.

One of my favorite experiments on the love hormone was performed on a large group of heterosexual men, who were all in monogamous relationships. Before the researchers introduced these men to an extremely attractive female stranger, they had them snort oxytocin into their noses through a medicinal nasal spray. The researchers gave the men in the experiment a few minutes for the oxytocin to form bonds with their oxytocin receptors. (Remember that oxytocin is a big peptide molecule, so it takes some time for it to get to its target location and bind with the receptors.) But once the researchers were confident that the bonds had been formed, the experiment could begin. They started by introducing the men one at a time to the beautiful woman. Then they monitored how closely the two people stood next to each other.

After collecting their data on the men in committed relationships, the researchers called in a group of single men. They repeated the experiment with the oxytocin nasal spray, and sent the single guys in one by one. Just like before, the researchers measured the physical distance between the men and the attractive stranger to see if they could determine the effect of the oxytocin molecules on the human body.

In general, they found that the men in relationships stood at least ten if not fifteen centimeters farther away from the attractive woman than the single men did. There were a few outliers, of course, but this study (among others) suggests that oxytocin in males leads to demonstrably stronger bonds between couples. So, the next time your hubby heads out to a bachelor party, squirt some oxytocin up his nose and give him a big kiss, and just let the chemistry do the rest.

If you have a super strong connection with your partner, you may be spending a lot of time together between the sheets. In these cases (and depending on your family planning strategy), you may want to take full advantage of the chemical reactions that are caused by a few other hormones, like levonorgestrel. In other words: birth control.

Levonorgestrel is a big molecule, technically a steroid hormone, that is often used in intrauterine devices (IUDs) to prevent pregnancy. The hormone is a close molecular cousin to testosterone, and when inside a women's body, it triggers two main chemical reactions. First, it triggers the production of a thick cervical mucus that actually obstructs the uterus from any invading sperm. And second, it causes the bonds to break within the uterine wall. This sheds the uterine lining and decreases its overall thickness, making it more difficult, if not impossible, for a fertilized egg to attach and grow. If that doesn't work, the IUD also *physically* prohibits the sperm from accessing any eggs accidentally released by the ovaries just by getting in the way of the flight path. Hence the famous T-shape. All three of these factors combine to give the hormonal IUD a one-year failure rate of 0.2%.

We can compare this to a copper IUD, which is the nonhormonal alternative to the one I just discussed. The copper IUD still maintains the plastic T-shape core, but instead of having levonorgestrel in the inside, it has copper wire wrapped around its outside. Once implanted, this device releases copper cations, which form bonds with the cervical mucus at the opening of the uterus, and triggers the production of a spermicidal molecule that attacks any sperm that attempts to enter the uterus.

The most popular form of contraception in the United States is oral birth control pills, which also use hormones to prevent pregnancy. Unlike IUDs (that can be effective for three to twelve years), birth control pills must be taken daily to provide a constant supply of estrogen and progestogen molecules to the women's body. So, every twenty-four hours, at the same time of day, the woman has to take her daily dose of hormones in order to maintain the same concentration of molecules in her bloodstream. With this distribution of hormones, the pill is able to trick your body into thinking it's pregnant, therefore it automatically stops ovulating when on the pill.

Luckily, scientists are aware of how difficult it is for a working adult to take the pill every day at the exact same time of day, so they found a way to adjust the concentration of the hormones in each dose to give us a little wiggle room (three hours). Because of human error, the pill is only effective at preventing pregnancy 91% of the time. For comparison, the condom has a success rate of 82% because it is simply latex (another polymer that can be generated from styrene, like the polystyrene coolers we discussed in the beach chapter) that physically blocks the entry of sperm into the uterus.

Regardless of the form you use, birth control is absolutely loaded with chemistry because of how the molecules provoke chemical reactions inside of the body. But what about the chemistry that happens outside of your body? That's where sex pheromones come in.

Pheromones are really big molecules released by one animal's body that affects the behavior of another animal's body. These substances were originally discovered back in 1959 by a German biochemist named Adolf Butenandt. This is the same scientist that had already received a Nobel Prize in Chemistry for synthesizing the first sex hormone just twenty years earlier. To describe him as the rock star of the chemistry world is an understatement.

His research led to the discovery that pheromones act like hormones, but just on a nearby member of the same species. For example, if animal A excretes sex pheromones near animal B, then the molecules will be absorbed into animal B's body and affect its overall behavior. This really means that animal A gets to act like Cupid, but instead of an arrow, he uses molecules.

For this reason, pheromones are sometimes referred to as eco-hormones, since they are molecules that operate like hormones outside of the body. And just like hormones, pheromones can have a variety of structures. Some of the molecules are very small, while others are quite big. They are all volatile molecules, which means that they can easily evaporate in a given set of conditions. We can usually detect volatile species because they are associated with a strong smell (like gasoline or nail polish remover).

Researchers decided to use the name pheromone because it means to transfer excitement, which is exactly what pheromones do. These powerful molecules can send signals to neighboring species about a number of different topics, like food, safety, or sex. For example, ants tell each other where food sources are by giving off pheromones on the path between their colony and the food. Dogs give off territorial pheromones when they pee on the fire hydrant to mark their territory during walks. Even male mice give off sexual pheromones that attract female mice, which then encourages neighboring male mice to become aggressive.

But what about humans? Do we give off any type of sex pheromones?

Contrary to popular belief, humans do not have any form of sex pheromone. But here's why everyone thinks we do: in 1986, Winnifred Cutler came out with research that claimed she had isolated the first human sex pheromone. In this project, she collected, froze, and then thawed what she reported were sex pheromones of a number of different people. A year later, she applied the molecule to the upper lip of many female subjects, after which she claimed that she observed similar results to what we see with animals out in nature.

Turns out, Cutler's research was a bunch of malarkey. She had not isolated a human sex pheromone; she was just putting weird smells on the upper lips of random test subjects. Including—get this—armpit sweat. Instead of isolating pure pheromones, she collected the electrolytes that we sweat out during perspiration and *put them on people's faces.*

To this day, Cutler's disgusting science is published all over the internet, which means some people can Google the human sex pheromone and get a lot of misinformation. Some researchers firmly believe that we will discover sex pheromones any day now, but at the time of the publication of this book, no human sex pheromone has been discovered. Many studies have been done, replicated, and adjusted for as many variables as possible, and each research group has come to the same conclusion: humans in the twenty-first century probably do not have a sex pheromone.

But has that always been the case? If most other mammals, like rabbits and goats, have sex pheromones, why don't we?

The answer is surprisingly simple. Humans learned how to communicate. We can use words (and candles…and lingerie…) to indicate to our partners that we are interested in knocking boots, while ferrets have to send sex molecules in the direction of their desired mate.

Before we leave the boudoir, there is one more hormone we must address: vasopressin, a large peptide molecule with multiple functions, like regulating our blood pressure and balancing our kidneys. During the arousal/excitement phase of the human sexual response cycle, the male body will release this hormone in conjunction with having an erectile response. After orgasm, the concentration of vasopressin in the bloodstream significantly decreases.

Vasopressin also plays a role in regulating the circadian rhythm, and because of this latter effect, it is theorized to trigger sleepiness and relaxation. We know that vasopressin levels surge highest in men during sexual activity...which might explain the near instantaneous, post-coital snooze.

But for women, the side effects from another hormone will kick in right about the same time as your head is coming out of the clouds. Melatonin is a hormone derived from amino acids that was discovered back in 1958 when an American chemist (later turned dermatologist) named Aaron Lerner was investigating treatments for skin diseases. While studying the glands of cattle, he stumbled upon melatonin, which is a relatively small molecule compared to the other hormones we have discussed in this section. This molecule is secreted from the pineal gland (located in the epithalamus in the center of our brains). The gland itself looks just like a pine cone—that's actually how it earned its name.

Then just twenty years later, Harry J Lynch's team at M.I.T. learned about melatonin's effect on the aforementioned human circadian rhythm and how it affected our sleep-wake cycle. If you are unfamiliar with circadian rhythms, they are essentially your body's version of a day planner. They dictate when chemical reactions happen within your body (like the ones associated with digestion or with sleep). For example, they are the primary reason why you feel hungry around 6:00 p.m., sleepy around 9:00 p.m., and then alert again around 7:00 or 8:00 a.m. the next morning. And why it's so damn difficult to work the night shift (or become a new parent).

Which brings us to our final, chemical laden activity of our day: sleep.

From a chemistry perspective, sleep is an altered state of consciousness, where the body processes through several chemical cycles. You probably already know that we alternate between rapid eye movement (REM) sleep and non-REM sleep. One cycle of REM and non-REM sleep takes about ninety minutes in humans, where the REM sessions increase in length as you continue to sleep.

REM sleep is not as tranquil as you think. During a typical REM cycle, your blood pressure increases, your heart starts to race, and your breath rate increases. More importantly, your brain is extremely active and produces tons of brain waves. When this happens, it's kind of like your brain is sorting through its daily mail. It throws away any useless memories (junk mail), and stores any important ones (bills). And all of this happens through movements of electrons within your brain.

Simultaneously, your muscles relax to the point of paralysis, which is ironic considering your brain is loaded with a large number of acetylcholine molecules. Why? When you are awake, these molecules are responsible for activating your muscles. However, when coupled with a lack of norepinephrine, serotonin, or histamine, your muscles freeze in order to allow for all of the body's energy to go into the chemical reactions happening in your brain.

But when we slip into a deeper sleep (before or after REM sleep), our body is completely shut off from the world. The GABA neurotransmitters (the same ones responsible for a person's drunken stupor) form bonds in the brain that inhibit its overall activity, which is why it is much more difficult to wake someone from non-REM sleep than REM sleep. Unfortunately, this is also when those who suffer from sleep disorders talk in their sleep or sleepwalk. And since our brain activity has been minimized during non-REM sleep, the things that come out of our mouths are usually complete and total gibberish.

My husband found out early in our relationship that I talk in my sleep and has been trying to record my late-night sagas ever since. Nine times out of ten, I'm mumbling about food or the dishes. But every once in a while, I give him an incoherent lesson on molecules and atoms.

I never seem to enunciate enough for him to decipher which topic I am teaching. Maybe I'm mumbling about the scattering of sunlight that causes crepuscular rays or how oxytocin and vasopressin are released during sex. Regardless, my tone and hand gestures are always clear. Even when I'm sleeping, I want the world to appreciate the chemistry behind our day-to-day lives.

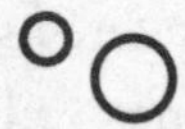

Have you ever heard the old fable about two fish swimming along a river? An older fish comes up to them and says something along the lines of "Good morning. How's the water?" After swimming along for a while, one of the two younger fish turns to the other to ask, "What's water?"

I love this little story because it perfectly encompasses the way that so many people experience chemistry in their lives. The truth is most adults leave chemistry behind in high school or college. Want proof? Only 3% of my students graduate as chemistry majors. When they walk off campus on the last day of school, they say goodbye and good riddance to my class about energy and matter. Yet, as I've hopefully proven to them (and to you), chemistry explains all the phenomena around us, all while helping us to understand the fabric of our reality.

We see it in the chemical reactions that help us block pain and digest food, the polymers used in our hair products and pies, the all-purpose cleaners used to wipe down our counters and bathrooms, and even in the last breath you just inhaled/exhaled. Chemistry touches every aspect of our lives. And if you look closely enough, you will see it in every scientific discipline, and every industry too—from clothing to makeup to toys and pharmaceuticals.

But as Carl Sagan said, "Science is a way of thinking much more than it is a body of knowledge." It's about asking *why* and *what if*, and then searching for answers until you reach the point of exhaustion. I hope this book inspires you to think critically about your environment, to keep educating yourselves, and to explore the wonders of the microscopic (and microcosmic) aspects all around us. I hope you stumble upon a topic that makes you feel the way I do about chemistry. And that you scream it from your rooftop until your neighbors beg you to stop.

Because when you are a nerd about something, a real, true, jump-up-and-down-in-the-chair-can't-control-yourself nerd about something, anything—and I mean *anything*—is possible.

★★★★★

Periodic Table of Elements

1A 1	2A 2	3B 3	4B 4	5B 5	6B 6	7B 7	8B 8	8B 9
1 **H** 1.008								
3 **Li** 6.941	4 **Be** 9.012							
11 **Na** 22.99	12 **Mg** 24.31							
19 **K** 39.10	20 **Ca** 40.08	21 **Sc** 44.96	22 **Ti** 47.87	23 **V** 50.94	24 **Cr** 52.00	25 **Mn** 54.94	26 **Fe** 55.85	27 **Co** 58.9
37 **Rb** 85.47	38 **Sr** 87.62	39 **Y** 88.91	40 **Zr** 91.22	41 **Nb** 92.91	42 **Mo** 95.94	43 **Tc** (98)	44 **Ru** 101.07	45 **Rh** 102.9
55 **Cs** 132.91	56 **Ba** 137.33	57 **La** 138.91	72 **Hf** 178.49	73 **Ta** 180.95	74 **W** 183.84	75 **Re** 186.21	76 **Os** 190.23	77 **Ir** 192.2
87 **Fr** (223)	88 **Ra** (226)	89 **Ac** (227)	104 **Rf** (261)	105 **Db** (262)	106 **Sg** (266)	107 **Bh** (264)	108 **Hs** (277)	109 **Mt** (268)

58 **Ce** 140.12	59 **Pr** 140.91	60 **Nd** 144.24	61 **Pm** (145)	62 **Sm** 150.36	63 **Eu** 151.96	64 **Gd** 157.2
90 **Th** 232.04	91 **Pa** 231.04	92 **U** 238.03	93 **Np** (237)	94 **Pu** (244)	95 **Am** (243)	96 **Cm** (247)

8B 10	1B 11	2B 12	3A 13	4A 14	5A 15	6A 16	7A 17	8A 18
								2 He 4.003
			5 B 10.81	6 C 12.01	7 N 14.01	8 O 16.00	9 F 19.00	10 Ne 20.18
			13 Al 26.98	14 Si 28.09	15 P 30.97	16 S 32.07	17 Cl 35.45	18 Ar 39.95
Ni 8.69	29 Cu 63.55	30 Zn 65.38	31 Ga 69.72	32 Ge 72.64	33 As 74.92	34 Se 78.96	35 Br 79.90	36 Kr 83.80
Pd 6.42	47 Ag 107.87	48 Cd 112.41	49 In 114.82	50 Sn 118.71	51 Sb 121.76	52 Te 127.60	53 I 126.90	54 Xe 131.29
Pt 5.08	79 Au 196.97	80 Hg 200.59	81 Tl 204.38	82 Pb 207.20	83 Bi 208.98	84 Po (209)	85 At (210)	86 Rn (222)
0 Ds 281)	111 Rg (281)	112 Cn (285)	113 Nh (286)	114 Fl (289)	115 Mc (289)	116 Lv (293)	117 Ts (293)	118 Og (294)

5 Tb 58.93	66 Dy 162.50	67 Ho 164.93	68 Er 167.26	69 Tm 168.93	70 Yb 173.04	71 Lu 174.97
7 Bk 247)	98 Cf (251)	99 Es (252)	100 Fm (257)	101 Md (258)	102 No (259)	103 Lr (262)

ACKNOWLEDGMENTS

I have to start by thanking Mary Jackson, the first Black female engineer at NASA. In moments of self-doubt, I think about your story. I focus on your tenacity and determination and how you refused to give up on your dream. Thank you for being a trailblazer and the best role model for women in STEM. I give you my word, right here and right now, that I will do everything in my power to make science a little easier for the next generation of girls, just like you did for me.

To my partner in crime and manager, Glenn Schwartz, thank you for contacting me back in January 2018. I still don't know what it was about your email that made me decide to take your call, but I'm so glad I did. You've given me a life I never knew could be possible. "Thank you" doesn't begin to express my appreciation.

I would like to thank the entire team at Park Row and HarperCollins for helping me make chemistry accessible (and fun)! Special thanks to my brilliant editor Erika Imranyi, who held my hand every step of the way. Erika, thank you for your patience, your kindness, and for pushing me to make this the best book possible. I've learned so much from you. Thanks for teaching me a "few" things about punctuation along the way.

And to Brandi Bowles and Meghan Stevenson, thank you for reading draft after draft after draft of this book. I appreciate every edit, suggestion, and critique you both gave me—thank you so much for making sure that this book didn't turn out like a lab report!

To the members of my DnD group (Jordan Corbman, Hannah Robus, Olin Robus, Dustin Myers, and Josh Biberdorf), thank you for keeping me sane and for working around my crazy schedule. Our weekly games bring Bimpnottin Loopmottin Waywocket Oda Orla Caramip Murnig Fnipper so much joy.

This book would not have been possible without the love and support of the following people: Craig and Teresa Crawford, Jack and Dort Crawford, Brendan and Daney Hughes, Brittany Crawford and Landon Hamilton, Katie and Becky Hughes, Caitlin Chambers, Chelsea Hoard, Kelsea Maul, Kathy and Smoz, Kim and Ivars Bergs, the Scroats, Kelli Palsrok, Kathleen Nolta, Vincent Pecoraro, John Wolfe, Alan Cowley, Simon Humphrey, David Vanden Bout, Paul McCord, Stacy Sparks, Jenny Brodbelt, Jen Moon, and Betty and Dort.

Lastly, to my mountain man, Josh Biberdorf. Thank you for your unconditional love and support. Thank you for making me dinner every night and bringing it to me while I worked on this book. Thank you for the back rubs, the flirtatious winks, and making me laugh every single day. More important, thanks for cheering me up on particularly rough days. You're my favorite person on this planet, and I can't wait to see what our next chapter holds. Muah.

SELECTED BIBLIOGRAPHY

Alberts, Bruce, Alexander Johnson, Julian Lewis, Martin Raff, Keith Roberts, and Peter Walter. *Molecular Biology of the Cell.* New York: Garland Science, 2002.

Atkins, Peter, and Loretta Jones. *Chemical Principles.* New York: W. H. Freeman and Company, 2005.

The American Chemical Society. *Flavor Chemistry of Wine and Other Alcoholic Beverages.* Portland: ACS Symposium Series eBooks, 2012. PDF e-book.

The American Chemical Society. *Chemistry in Context.* New York: McGraw-Hill Education, 2018.

The American Chemical Society. *Flavor Chemistry of Wine and Other Alcoholic Beverages.* United Kingdom: OUP USA, 2012.

Aust, Louise B. *Cosmetic Claims Substantiation.* New York: Marcel Dekker, 1998.

Barel, André, Marc Paye, and Howard I. Maibach, ed. *Handbook of Cosmetic Science and Technology.* Boca Raton: Taylor & Francis Group, 2010.

Barth, Roger. *The Chemistry of Beer.* Hoboken: John Wiley & Sons, Inc., 2013.

Belitz, Hans-Dieter, Werner Grosch, and Peter Schieberle. *Food Chemistry.* Berlin Heidelberg: Springer-Verlag, 2009.

Beranbaum, Rose Levy. *The Pie and Pastry Bible.* New York: Scribner, 1998.

Black, Roderick E., Fred J. Hurley, and Donald C Havery. "Occurrence of 1,4-dioxane in cosmetic raw materials and finished cosmetic products." *Journal of AOAC International 84*, no. 3 (May 2001): 666–670.

Bouillon, Claude, and John Wilkinson. *The Science of Hair Care.* Abingdon: Taylor & Francis, 2005.

Boyle, Robert. *The Sceptical Chymist.* London: J. Cadwell, 1661.

Crabtree, Robert H. *The Organometallic Chemistry of the Transition Metals.* Hoboken: Wiley-Interscience, 2005.

The Editors of *Cook's Illustrated. The New Best Recipe.* Brookline: America's Test Kitchen, 2004.

Eğe, Seyhan. *Organic Chemistry.* Boston: Houghton Mifflin Company, 2004.

Feyrer, James, Dimitra Politi, and David N. Weil. "The Cognitive Effects of Micronutrient Deficiency: Evidence from Salt Iodization in the United States." *Journal of the European Economic Association 15*, no. 2 (April 2017): 355–387.

"Foundations of Polymer Science: Wallace Carothers and the Development of Nylon." American Chemical Society National Historic Chemical Landmarks. American Chemical Society. Accessed March 12, 2020. http://www.acs.org/content/acs/en/education/whatischemistry/landmarks/carotherspolymers.html.

Fromer, Leonard. "Prevention of anaphylaxis: the role of the epinephrine auto-injector." *The American Journal of Medicine 129*, no. 12 (August 2016): 1244–1250.

Fuss, Johannes, Jörg Steinle, Laura Bindila, Matthias K. Auer, Hartmut Kirchherr, Beat Lutz, and Peter Gass. "A runner's high depends on cannabinoid receptors in mice." *PNAS 112*, no. 42 (October 2015): 13105–13108.

"Gchem." McCord, Paul, David Vanden Bout, and Cynthia LaBrake. The University of Texas. Accessed December 20, 2019. https://gchem.cm.utexas.edu/.

Goodfellow S.J., and W.L. Brown. "Fate of Salmonella Inoculated into Beef for Cooking." *Journal of Food Protection 41*, no. 8 (August 1978): 598–605.

Green, John, and Hank Green. Vlogbrothers' YouTube page. Accessed May 15, 2020. https://youtu.be/rMweXVWB918?t=75.

Guinn, Denise. *Essentials of General, Organic, and Biochemistry.* New York: W. H. Freeman and Company, 2014.

Halliday, David, Robert Resnick, and Jearl Walker. *Fundamentals of Physics.* Hoboken: John Wiley & Sons, Inc., 2014.

Hammack, Bill, and Don DeCoste. *Michael Faraday's The Chemical History of a Candle with Guides to the Lectures, Teaching Guides & Student Activities.* United States: Articulate Noise Books, 2016.

Higginbotham, Victoria. "Copper Intrauterine Device (IUD)." *Embryo Project Encyclopedia* (July 2018): 1940–5030.

Hodson, Greg, Eric Wilkes, Sara Azevedo, and Tony Battaglene. "Methanol in wine." *40th BIO Web of Conferences 9*, no. 02028 (January 2017): 1–5.

Horton, H. Robert, Laurence A. Moran, Raymond S. Ochs, J. David Rawn, and K. Gray Scrimgeour. *Principles of Biochemistry.* Upper Saddle River: Prentice Hall, Inc., 2002.

Housecroft, Catherine E., and Alan G. Sharpe. *Inorganic Chemistry.* Harlow: Pearson, 2018.

"How Big Is a Mole? (Not the animal, the other one.)" Daniel Dulek. TED Talk. Accessed August 3, 2019. https://www.ted.com/talks/daniel_dulek_how_big_is_a_mole_not_the_animal_the_other_one/transcript?language=en.

Iizuka, Hajime. "Epidermal turnover time." *Journal of Dermatological Science 8*, no. 3 (December 1993): 215–217. https://linkinghub.elsevier.com/retrieve/pii/0923181194900574.

Karaman, Rafik. *Commonly Used Drugs: Uses, Side Effects, Bioavailability and Approaches to Improve It.* United States: Nova Science Incorporated, 2015.

King Arthur Flour. *The All-Purpose Baking Cookbook.* New York: The Countryman Press, 2003.

Koltzenburg, Sebastian, Michael Maskos, and Oskar Nuyken. *Polymer Chemistry.* Berlin Heidelberg: Springer-Verlag, 2017.

Lynch, Harry J., Richard J. Wurtman, Michael A. Moskowitz, Michael C. Archer, and M.H. Ho. "Daily rhythm in human urinary melatonin." *Science 187*, no. 4172 (January 1975): 169–171.

"Making sense of our senses." Maxmen, Amy. Science. Accessed February 2020. https://www.sciencemag.org/features/2013/11/making-sense-our-senses.

Marks, Lara. *Sexual Chemistry.* New Haven, London: Yale University Press, 2010.

McGee, Harold. *On Food and Cooking.* New York: Scribner, 2004.

Moberg, Kerstin Uvnäs. *The Oxytocin Factor.* London: Pinter & Martin, 2011.

Nehlig, Astrid, Jean-Luc Daval, and Gerard Debry. "Caffeine and the central nervous system: mechanisms of action, biochemical, metabolic and psychostimulant effects." *Brain Research Reviews 17*, no. 2 (May 1992): 139–170.

Norman, Anthony W., and Gerald Litwack. *Hormones.* San Diego, California: Academic Press, 1997.

"Nylon: A Revolution in Textiles." Audra J. Wolfe. Science History Institute. Accessed March 14, 2020. http://sciencehistory.org/distillations/magazine/nylon-a-revolution-in-textiles.

O'Lenick, Anthony J., and Thomas G. O'Lenick. *Organic Chemistry for Cosmetic Chemists.* Carol Stream: Allured Publishing, 2008.

Oxtoby, David W., H.P. Gillis, and Alan Campion. *Principles of Modern Chemistry.* Belmont: Brooks/Cole, 2012.

"Parabens in Cosmetics." U.S. Food & Drug Administration. Accessed September 14, 2019. https://www.fda.gov/cosmetics/cosmetic-ingredients/parabens-cosmetics.

Partington, James Riddick. *A Short History of Chemistry.* New York: Dover Publications, 1989.

"Periodic Table of Elements." International Union of Pure and Applied Chemistry. Accessed October 20, 2019. https://iupac.org/what-we-do/periodic-table-of-elements/.

"Pheromones Discovered in Humans." Boyce Rensberger. Athena Institute. Accessed March 3, 2020. http://athenainstitute.com/mediaarticles/washpost.html.

Richards, Ellen H. *The Chemistry of Cooking and Cleaning.* Boston: Estes & Lauriat, 1882.

Roach, Mary. *Bonk: The Curious Coupling of Science and Sex.* New York, London: W. W. Norton & Company, 2008.

Robbins, Clarence R. *Chemical and Physical Behavior of Human Hair.* New York: Springer Science+Business Media, LLC, 1994.

Sakamoto, Kazutami, Robert Y. Lochhead, Howard I. Maibach, and Yuji Yamashita. *Cosmetic Science and Technology.* Amsterdam: Elsevier Inc., 2017.

Scheele, Dirk, Nadine Striepens, Onur Güntürkün, Sandra Deutschländer, Wolfgang Maier, Keith M. Kendrick, and René Hurlemann. "Oxytocin modulates social distance between males and females." *Journal of Neuroscience 32*, no. 46 (November 2012): 16074–16079.

Scheer, Roddy, and Doug Moss. "Should People Be Concerned about Parabens in Beauty Products?" *Scientific American*, October 2014, https://www.scientificamerican.com/article/should-people-be-concerned-about-parabens-in-beauty-products/.

Simons, Keith J., and F. Estelle R. Simons. "Epinephrine and its use in anaphylaxis: current issues." *Current Opinion in Allergy and Clinical Immunology 10*, no. 4 (August 2010): 354–361.

Smith, K.R., and Diane Thiboutot. "Sebaceous gland lipids: friend or foe?" *Journal of Lipid Research 4* (November 2007): 271–281.

Spellman, Frank R. *The Handbook of Meteorology.* Plymouth: Scarecrow Press, Inc., 2013.

Spriet, Lawrence L. "New Insights into the Interaction of Carbohydrate and Fat Metabolism During Exercise." *Sports Medicine 44*, no. 1 (May 2014): 87–96.

Society of Dairy Technology. *Cleaning-in-Place: Dairy, Food and Beverage Operations.* Oxford: Blackwell Publishing, 2008.

Srinivasan, Shraddha, Kriti Kumari Dubey, Rekha Singhal. "Influence of food commodities on hangover based on alcohol dehydrogenase and aldehyde dehydrogenase activities." *Current Research in Food Science 1* (November 2019): 8–16.

"Sunscreens and Photoprotection." Gabros, Sarah, Trevor A. Nessel, and Patrick M. Zito. StatPearls Publishing. Accessed January 15, 2020. https://www.ncbi.nlm.nih.gov/books/NBK537164/.

Tamminen, Terry. *The Ultimate Guide to Pool Maintenance.* New York: McGraw-Hill Education, 2007.

The Royal Society of Chemistry. *Coffee.* Croydon: CPI Group (UK), 2019.

"This 16-year-old football player lifted a car to save his trapped neighbor." Ebrahimji, Alisha. CNN. Accessed January 19, 2020. http://cnn.com/2019/09/26/us/teen-saves-neighbor-car-trnd/index.html.

Toedt, John, Darrell Koza, and Kathleen Van Cleef-Toedt. *Chemical Composition of Everyday Products.* Westport: Greenwood Press, 2005.

Tosti, Antonella, and Bianca Maria Piraccini. *Diagnosis and Treatment of Hair Disorders.* Abingdon: Taylor & Francis, 2006.

Tro, Nivaldo J. *Chemistry.* Boston: Pearson, 2017.

Waterhouse, Andrew Leo, Gavin L. Sacks, and David W. Jeffery. *Understanding Wine Chemistry.* Chichester: John Wiley & Sons, Inc., 2016.

Wermuth, Camille Georges, David Aldous, Pierre Raboisson, Didier Rognan, ed. *The Practice of Medicinal Chemistry.* London, England: Academic Press, 2015.

Young, David, John D. Cutnell, Kenneth W. Johnson and Shane Stadler. *Physics.* Hoboken: John Wiley & Sons, Inc., 2015.

"Your Guide to Physical Activity and Your Heart." National Institutes of Health, National Heart, Lung, and Blood Institute. Accessed March 23, 2020. http://nhlbi.nih.gov/files/docs/public/heart/phy_activ.pdf.

Zakhari, Samir. "Overview: How is Alcohol Metabolized by the Body?" *Alcohol Research & Health 29*, no. 4 (2006): 245–254.

Zumdahl, Steven S. *Chemical Principles.* Belmont: Brooks/Cole, 2009.

Zumdahl, Steven S., Susan A. Zumdahl, and Donald J. DeCoste. *Chemistry.* Boston: Cengage Learning, 2018.

GLOSSARY

Acid: a molecule with a pH lower than 7

Aerobic: a reaction that needs oxygen to occur

Alcohol: molecules (usually hydrocarbons) that contain an oxygen-hydrogen covalent bond

Amino acids: molecules that contain only carbon, hydrogen, nitrogen, and oxygen, atoms that are necessary for human life

Anaerobic: a process that occurs without the presence of oxygen

Anion: a negatively charged atom

Aromatic: molecules that are fragrant in nature

Atom: the fundamental building block of matter (contains protons, neutrons, and electrons)

Atomic mass: the sum of the protons and weighted average of neutrons in an atom

Atomic number: the number of protons in an atom

Base: a molecule with a pH greater than 7

Bond: a chemical interaction between two atoms (usually by sharing or transferring electrons)

Carbohydrates: the sugar and starch molecules in our foods

Catalyst: a molecule that provides an alternate pathway for a chemical reaction (and usually increases the rate of reaction)

Cation: a positively charged atom

Cis: the orientation that occurs when both functional groups are on the same side of the molecule

Covalent bond: an interaction that occurs when two atoms share electrons

Density: the relative mass occupied by a substance in a specific volume

Dipole-dipole: IMFs that occur between two polar molecules

Dispersion forces: IMFs that occur between two nonpolar molecules

Electrolytes: ionic species (or salts)

Electromagnetic radiation: electromagnetic waves that propagate through space in the form of radio, microwave, infrared, visible, ultraviolet, X-ray, and gamma radiation

Electron: a negatively charged particle located outside of the nucleus of an atom

Electronegativity: a measure of how attracted one atom's electrons are to another atom's nucleus

Element: a collection of atoms with the same number of protons (and physical/chemical properties)

Endothermic: a process that absorbs energy (becomes colder)

Enzymes: naturally occurring molecules that act like catalysts to cause a chemical reaction (often within the human body)

Exothermic: a process that releases energy (becomes warmer)

Fatty acids: a long molecule that has a nonpolar end (hydrocarbons) and a polar end (a carboxylic acid)

Functional groups: one part of the molecule that greatly affects the chemical reactivity of the entire molecule

Glucose: a monosaccharide (sugar) with the molecular formula $C_6H_{12}O_6$

Hormone: a molecule that carries "messages" from one place to another in the body

Hydrocarbon: a molecule that contains only hydrogen and carbon atoms

Hydrogen bonding: IMFs that occur between two molecules that each contain covalent bonds between hydrogen and either nitrogen, oxygen, or fluorine atoms

Hydrophobic: a nonpolar molecule that repels water

Intermolecular forces (IMFs): attractions that occur between molecules

Intramolecular forces: attractions within the molecule (usually bonds between atoms)

Ion: a charged atom (it can be positive or negative)

Ionic bond: an interaction that occurs when one atom transfers electrons to another atom

Isotopes: two or more elements that have the same number of protons, but a different number of neutrons

Macroscopic: something that can be observed with the human eye (without special instruments)

Mass number: the number of protons and neutrons in an atom

Microscopic: something that cannot be observed with the human eye (without special instruments)

Molecule: a substance that contains two or more atoms

Neutron: a neutrally charged particle located in the nucleus of an atom

Nonpolar: a molecule (or bond) that has an even distribution of electrons

Nucleus: the center of the atom (contains protons and neutrons)

Peptide: a molecule made of two or more amino acids

Polar: a molecule (or bond) that has an uneven distribution of electrons

Polymer: large molecules of repeating units

Polypeptides: the protein molecules in our foods

Proton: a positively charged particle located in the nucleus of an atom

Thermal energy: kinetic energy in the form of heat

Trans: the orientation that occurs when both functional groups are on the opposite side of the molecule

Triglycerides: the molecules in fats and oils in our foods

Valence electron: electrons in the outer layer of the atom

Vaporization: the phase change that occurs when a liquid changes into a gas